PYGMY GOATS

MANAGEMENT AND
VETERINARY CARE

Other books available from All Publishing Company:

The Illustrated Standard of the Pygmy Goat
written by Lorrie Blackburn, D.V.M., illustrated by
Ann Werts, and edited by Lydia Hale

Pot-Bellied Pet Pigs, Mini-Pig Care and Training
by Kayla Mull and Lorrie Blackburn, D.V.M.

Veterinary Care of Pot-Bellied Pet Pigs
by Lorrie Boldrick, D.V.M.
(a.k.a. Lorrie Blackburn, D.V.M.)

Front cover photograph by Lydia Hale

PYGMY GOATS

MANAGEMENT AND
VETERINARY CARE

BY

LORRIE BOLDRICK, D.V.M.
A.K.A. LORRIE BLACKBURN, D.V.M.

AND

LYDIA HALE

All Publishing Company
Orange, California
1996

Published by: All Publishing Company
10951 Meads Avenue
Orange, CA 92869

Copyright © 1996 by Lorrie Boldrick, D.V.M.
All Rights Reserved
Printed in the United States of America
August 1996 First Edition

Library of Congress Catalog Card Number: 96-85589

ISBN 0-962-4531-3-7

TABLE OF CONTENTS

PREFACE

This book has been a long-time undertaking, and the pages herein are a compilation of experiences and expertise gained over twenty-five years of owning and caring for pygmy goats. The enclosed information is not intended to be all inclusive nor the final word on pygmy goats, but rather a general guide and reference. A Glossary in the back is for ready referral.

Special thanks to Ann Werts for giving permission to reprint illustrations, some of which appeared in previous publications, and for contributing new ones for us. Our gratitude also to Allie Blackburn for her cartoons and to Jane Bryant for her excellent technical editing. And thank you to Dr. Elizabeth Buchholz, Diane Siebert and Connie Sweet for reading early copies of this manuscript and making suggestions for improvement and clarification.

We hope this book will be helpful to the pygmy goat hobbiest and the veterinarian alike.

Lorrie Boldrick and Lydia Hale

INTRODUCTION

WHY GOATS?

by Ann Werts

Over the years, the one question I have invariably been asked is "Why goats?" Everyone from new acquaintances to old friends ask this question sooner or later. Early in my goat keeping career the question even puzzled me. I would pause and reflect. Why goats? My answer would then be a long narrative on the endearing qualities of the little caprus. As years went by and *that* question became expected rather than surprising, my answer became less complicated and to the point. They are wonderful and sensitive and I love them - leaving the examiner to ponder my sanity or to accept my unorthodox hobby.

Recently when I was attending a seminar this question was asked by a woman who no doubt had never been exposed to the joys of animals or of goat keeping. She was polished from head to toe. Every hair on her head was in place and her acrylic fingernails were at least 3 inches long and painted bright red. She was impeccably dressed from her Gucci bag to her Dior suit. There was not a single animal hair anywhere on her. I knew immediately she could never pull a kid.

She asked "Why goats?" with an air of superiority and a slight sneer. My response was the short "I love them" version which was accepted with a look of distrust or disdain, I am not sure which.

1

Later when I was pondering this incident I decided I would research a more educated answer rather than the simplistic "I love them" response. It occurred to me that I really did love them, but why. After two years of research, I have a better understanding of this innate love I feel.

Goats have been around a very long time, emerging in the geological record five million years ago in late Pliocene times. They thrived in rocky, infertile regions, which in time put a premium on intelligence, agility and ability to forage widely in small, self-reliant groups. Goats gradually spread out to occupy all regions of the Old World. Their domestication is now dated to the middle of the eighth millennium B.C. Bones of goats dating as far back as 10,000 B.C. were found in the Middle East at Al Khaim in the Wadi Kharaitun which runs from Bethlehem to the Dead Sea.

Osteologists have also determined that the goat was the most popular and plentiful of the early domesticated animals. Eighty percent of all animal bones excavated in the early Neolithic settlements of 7000 B.C. such as Jarmo in the Middle East, were of domesticated goats. This same discovery was made in the earliest agricultural settlements of Thessaly, Greece.

Other archaeologists claim that the oldest authentic sign of animal husbandry was found in the lower strata of the city of Jericho and also near Ur on the Euphrates, the area in the Middle East know as "the cradle of man." Here the bones of goats were estimated to be seven thousand years old.

It has been established that even before they practiced agriculture, the ancient Egyptians kept herds of goats. One of their most respected Gods was the goat Mendes. The Pharaoh Cephrenes thought so highly of goats that he had 2234 entombed with him near the pyramid of El Gezeh. Egyptian art portrays goats being herded over the Nile-damaged fields to trample seeds into the soil so the birds couldn't steal them. Many Egyptian artifacts and tomb paintings depict goats. On the prow of an ancient Egyptian toy boat was carved a horned goat. A bas-relief from an ancient Egyptian tomb depicts goats nibbling on trees and bushes.

By the third millennium B.C., the Sumerians considered goats so valuable that they could be exchanged for silver or copper. One of the most ancient motifs in Sumerian art is that of a buck standing on its hind legs in the thicket, browsing on the Tree of Life.

With its strong features and air of dignity, the historic goat was the first animal to appear in ancient Greek and Roman mythology. It was a face to stir the imagination. The gods of the Greeks, like many less important mortal babies, were nourished on goat's milk. Old Zeus, born in a Cretan cave, was brought up by the doe Amalthia; when he became king of heaven, he rewarded his nurse by placing her in the sky, turning her into the bright star Capella ("Little Goat"). Dionysis, the God of wine, was another suckled on goat's milk. Pan was the god of the country and fertility and protected flocks. He is depicted as having horns and goat's legs. The syrinx or pipes were said to have been invented by Pan. If you heard his pipes,

3

there was no saying what might happen. One of the Homeric Hymns is dedicated to Pan:

> Tell me, Muse, concerning the dear son of Hermes, the goat-footed, the two-horned, the lover of the din of revel, who haunts the wooded dells with dancing nymphs that tread the crest of the steep cliffs, calling upon Pan the pastoral God of the long witch hair.

Along with certain godly powers, Pan had the mischievous playful disposition of a goat. He liked to jump out and startle lonely travelers in the wilder places. And no doubt he was blamed for spoiling many a holiday in the country, turning a picnic into a Pan-ic. That is how the word panic derived. The definition is unreasonable fright, and perhaps a stampede such as flocks of goats make at times for no apparent cause. Pan is even said to have gained several military victories for the Greeks by panicking and scaring off attacking armies. The enemy thought he was "the Devil himself," also often pictured with horns, goatee, and cloven hooves.

"Tragedy" comes from the Greed word for "goat." The Greeks also seem to have had a dance called the goat dance. This dance is depicted on a Greek vase which is in the British Museum.

The goat may also have been the first creature to have its name written in the heavens. Capricornus, the goat, has for centuries been the tenth sign of the Zodiac. Capricorn is a major constellation in the southern skies and the Tropic of Capricorn marks the maximum southern deviation of the sun during the year.

In ancient Viking mythology Thor, the goat of thunder, war, and strength used goats to pull his chariot across the skies. In the myth "The Stealing of Thor's Hammer" Thor had a dream that a thief crept into his bedroom and stole the one sure protection the gods had against the giants - his hammer. When he awoke, to his disdain the dream had been true. Thor summoned the god Loki to help him recover the hammer. Thor and Loki concocted an elaborate plan. But in order to reach the planned destination, Thor needed to use his chariot. Thor's goats were led from the stable and harnessed to the chariot. "Come on there, Toothgnasher! Go, Toothgrinder!" he shouted and cracked his whip. In a flash of lightning they were halfway across the sky. It is interesting to note the similarity to our own myth or story of Santa Claus' sled being pulled by reindeer, a close relative of the goat. The names Toothgnasher and Toothgrinder are also interesting. Who has not heard their own goats grinding and gnashing their teeth as they chew their cuds.

The Phoenicians were probably the first to bring goats to the ancient Celtic people of the British Isles. The Phoenicians were great ship builders, navigators, and merchants of their times. They are reported to have traded goats for the metal tin.

The ancient Celtic priests, the doctors of their day, dealt as much in magic and spirits as in medicine. They developed prescriptions for healing the sick by use of goats or portions of goats. Goats were thought useful in curing "falling sickness," perhaps because they never get dizzy. First the priest would apply a dog's gall to the patient's head. Then he would burn a goat's horn blowing all the foul smelling, suffocating smoke at the patient, who it was claimed would never have dizzy spells again.

To cure the loss of sight, the Celtic priest would boil a goat's liver and the patient would keep his eyes open in the rising steam. To cure deafness, the priest dropped melted and salted goat's fat in a person's ears. The burning of goat's hair was supposed to dispatch marauding serpents. Ashes from goat's charred hip bones were considered a good toothache powder. Goat's blood was used as a cure for poison, for the bites of "creeping worms" and scorpions, and also for indigestion and other such internal ailments. It was also reported as fine for making a person perspire.

The goat has survived in the Bible, where it is mentioned 136 times. The work "kid" is used 51 times and there are 11 direct references made to goat's milk, hair, and skin. The Old Testament points out there worth, mentioning that Jacob received goats as wages.

"And before him shall be gathered all nations, and he shall separate them one from another as a shepherd divideth his sheep from goats. And he shall set the sheep on his right hand, but the goats on the left." Church fathers took this literally and assigned the followers of Christ to sheep and banished the goat as a pagan thing.

It was also the Scriptures that gave us a word we use today - "scapegoat," or fall guy, one who is blamed for another's wrongdoing. In ancient times, in various places, pagan people believed that every so often they must offer something of great value to terrible gods that might destroy their tribe. Goats represented wealth, so they would trade a goat's precious life to preserve their own. On a certain day each year, all the sins of a whole community would be confessed over a goat; and then the animal would be taken out and lost in the wilderness. It was believed that it carried away all the confessed sins with it. Thus the sinners escaped or escaped punishment. Even today, in underdeveloped countries, to end a epidemic or plague, all sins are confessed over a goat and it is banished, the people praying that the plague will go away.

During the Middle Ages, the goat was thought to be one of the familiars of witches and by the 19th century, the Devil was portrayed as a black goat with glowing eyes. In Medieval times there was no more powerful and persistent symbol of the forces of evil, of the Devil himself, than the goat.

The votaries of the Dianic cult were rumored to possess the power to turn human beings into animals. At least one survival of this belief seems to have cropped up in medieval tales about witches. A sculptured stone panel at the entrance of Lyons Cathedral pictures a witch riding a man she has changed into a goat. She is whirling a cat around her unfortunate victims head, so that it may tear his face with its claws and is apparently on her way to a Sabbat meeting. It is reported that time after time in witch trials, the accused would assert that the Devil appeared to them in the form of a great black goat.

As stated before, the goat was considered in pagan

times to be a symbol of fertility. Half human and half bestial with horns and cloven hooves, he appeared as Dionysus or Bacchus, the chief fertility god of the classical world and was also found in the pantheons of northern Europe. Apparently, the early Christians thought him the most abominable of all the pagan deities; they gave his attributes, his horns and cloven hooves, to the Devil, adding to those the wings of the fallen angel.

The differences between the goat as fertility god and the goat as Devil or evil spirit sprang from the different attitudes of pagans and early Christians as mentioned before.

Some modern students of witchcraft claim that the witches really did worship a goat-god, not the Devil who is a Judeo-Christian creation, but rather an ancient horned deity. Other scholars believe that the witch-hunters simply assumed that witches worshiped the Devil, horns, hooves, and all, and that many of those accused of witchcraft were forced to confess to dealings with such a creature.

However the goat got its reputation as a symbol of the Devil or as the Devil himself, it is the animal most commonly pictured and described in connection with witchcraft ceremonies. In the "Encyclopedia of Witchcraft and Demonology", Russell Hope Robbins quotes a typical confession made by an accused witch in 1594. The suspect's lover took her to a Sabbat on the eye of St. John the Baptist's Day (June 24). He made a magic circle and invoked a large black goat, two women, and a man dressed like a priest. When the man told the Devil-goat that the girl wished to become his subject:

> The goat ordered her to make the sign of the cross with her left hand and all to venerate him. At which all kissed him under the tail. Those present lit candles they were holding from a black candle burning between the goat's horns and dropped money in an offertory bowl.

It is reported that Columbus's ships, the Nina, Pinta, and Santa Maria carried goats to the new world. They were aboard to supply fresh milk to the sailors.

When the first settlers came from Europe to America, they brought with them a great many goats. They knew that goats could forage more ably than cows along fringes of the unbroken wilderness. As the land was cleared, forests giving way to grass and grain, more and more cows were introduced in the New World. Much of the success of the colonists was due in part to the tenacious, loving and ever-giving goat.

In modern America goats are used in many differing ways. Goats have been used for brush control in fire areas. Goat milk is often used in hospitals for ailing patients who cannot digest cow milk. The Dairy goat industry is a multi-million dollar business. Goats can be found living in many a horse stable throughout the U.S. It is there as a playful companion for the horse. The U.S. Navy's mascot is the goat and many a vessel today carries a goat for a mascot. A recent Cornell University study revealed that a goat would make the best companion in space.

Goats are also remembered kindly in literature and legends. It was a goat that was Robinson Crusoe's friend and salvation on his deserted island. In the classic Swiss tale, Heidi's grandfather found comfort in his goats. Columbus took a small herd of goats for fresh milk on his trip across the sea. John Barth wrote a fantastic novel titled "Giles Goat-Boy" in which a boy is raised as a goat. And who could forget the tale of "Billy Goat Gruff"?

It would seem that human-kind has shared its beginnings with the goat. It is no wonder to me that I've always had an innate feeling of love and respect for goats, and now I can truly understand how these creatures have always been a part of me.

I am not sure that next time I am asked "Why goats?" I will respond with this short version of the entire history of civilization or "because I love them," but I feel secure in the knowledge of "Why goats?"

History

The Pygmy Goat, as we know it in the United States, originated in the what was called the French Cameroon area of West Africa. The goat in its pure form was gray in color with black stockings and was very small in size. Similar forms of goats occurred in all of northern Africa in 1959 when the original goats were imported to California and New York.

Hoofstock could not be imported directly from Africa because of restrictions against hoof and mouth disease. There were no restrictions against importation into European countries, and goats were sent to many zoos in Europe. In 1959 Sweden was removed from the restricted list. Twelve or thirteen goats were in the original shipment to the United States from Sweden. With quarantine costs included, the original goats cost approximately $3500 each.

Early in the 1970s, there were already many herds being used in teaching and research at various colleges and universities. Interest in the Pygmy Goat began to grow in the private sector as well. Eventually a registry was established to define breed type and to encourage the breeding of pure pygmies.

Today, there are thousands of pygmy goats in private homes in the United States. There are several registries and pygmy goat organizations. Numerous pygmy goat shows are held yearly. Pygmy goats are considered to be livestock, but many of them are held strictly as pets. The Pygmy Goat is here to stay.

THE
NORMAL
PYGMY
GOAT

Description

The pygmy goat is small and compact with a distinct appearance unique to the strain. It is unlike the larger dairy goats not only in its diminutive size but also in its conformation, colors and markings. A good, purebred pygmy goat cannot be confused with any other breed.

Adult pygmy goats measure between 16 and 23 inches at the withers and weigh between 60 and 80 pounds, depending on sex. Their full-barreled bodies are set squarely on short, muscular legs set wide apart giving them a sturdy, cobby appearance. The head is short with a dished face and good width between the eyes. The ears are relatively short and erect. The muzzle is broad.

While many people prefer to leave the horns on their goats, others prefer to disbud young kids a few weeks after birth. Either is acceptable for registration, but genetic hornlessness is considered a disqualifying fault.

The hair coat of pygmy does and bucks is visibly different. Does have a short to medium-long smooth coat, while the bucks' coat is longer and more dense particularly over the withers where it forms a cape or martingale. The coats on both does and bucks may vary depending on the climate and area of the country in which they live. In the warmer climates, their coats will remain pretty much the same year round. In the northern colder areas, the goats grow a short thick undercoat which they will shed in the springtime. The hair itself, on both sexes, should be straight and rather coarse. Does may have no beard at all or a sparse one, whereas the bucks' beard is long and thick.

Colors of pygmy goats range from solid black to a creamy white (referred to as "caramel"), with varying shades in between. The most typical color is gray agouti which is produced by the intermingling of dark and light hairs resulting in a salt and pepper appearance. Gray agouti colors vary from a very light silver to a dark charcoal. The caramel colors range from white to a dark reddish tan.

Pygmy goats, regardless of the color, have distinct markings which are their trademark. They are typically marked with a dark facial mask, dark socks on all four legs, a dark dorsal stripe, and martingale (on bucks.) The ears, muzzle, forehead, and eyes are accented in a lighter frosted tone. The dark socks on the agouti goats are solid from the knees and hocks down. The caramels have a light vertical stripe on the front of each cannon. Solid black pygmies are the only exception and require no marking at all.

Random markings such as a belt or partial belt are fairly common and are acceptable if they are within the girth area (behind the shoulder and in front of stifle). Light or white patches in other areas are faulted.

Figure 1. Agouti color

The agouti color is created by the intermingling of light and dark hair. Agouti colored animals must have darker markings (usually solid black) below the knee and below the hock. They most often have a darker mask over part of the face. Lighter or white accented markings are required on the ears, muzzle, forehead and around the eyes.

Figure 2. Caramel color

The caramel body color can vary from white to a dark tan or brown. These animals are also required to have dark leg markings, but a light vertical stripe is allowed on the front of each cannon. In fact, very few caramel goats will have a solid dark stocking. The facial mask on does will usually be broken and will appear as two vertical lines just to the inside of the eyes. A dark dorsal stripe is also preferred.

Caramel bucks, in addition to the dark markings described for the doe, have a dark martingale over the shoulders. This involves the longer hairs of the mane and creates a very striking appearance.

Parts of the Goat

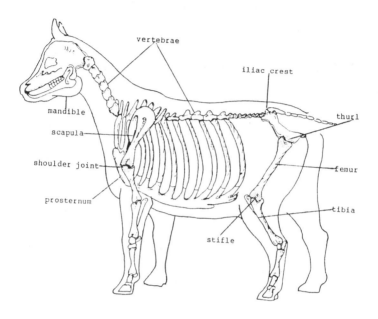

Figure 3. Skeleton

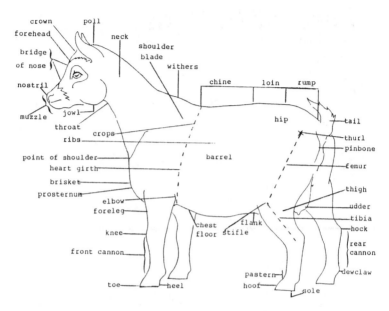

Figure 4. Parts of the Goat, side view

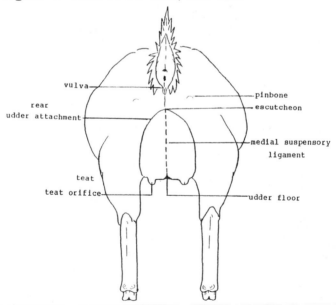

Figure 5. Parts of the Goat, rear view

20

Your Normal Animal

Sick animals always seem to be most important to the goat owner. But, how do you know if your goat is sick? The good herd manager knows his healthy animals so well that any subtle changes in behavior will alert him to the earliest visible signs of illness. He can then follow through with a more thorough evaluation of the animal's condition to decide if professional help or treatment is necessary.

The key to this is knowing your goats. This doesn't mean knowing the normal values of respiratory and pulse rates, although they will be discussed shortly. This means knowing your goat's normal behavior patterns, attitudes, posture, etc. If your goat usually knocks you down for her grain and today she doesn't - why? If she usually screams for hay the minute you walk out your back door and today she doesn't - why? If she's usually quiet and docile and today she is bleating and chasing the other does - why? If the one wild doe in your herd suddenly lets you pet her - why? These questions may all have answers that are unrelated to health problems. But these questions should come to mind. None of these situations should be ignored or left unanswered.

If you try to answer the above questions and don't come up with simple answers like "your daughter just fed the grain ten minutes earlier" or "she's in season" then you need to check the animal over more carefully. First, stand back and observe. Are there any other signs you should notice? Is the doe coughing, straining, preferring to lie down, pacing, circling, etc? Is she shivering or is her hair coat "puffed" up? Is she eating hay, eating grain, or chewing her cud? (Check this before catching her, while she's still relaxed.) Now catch her and check her temperature and respiration (pulse is difficult without experience). Check for rumination. Count the number of rumen contractions per minute and observe the strength of each contraction. Roll an eyelid out slightly and look at the color of the mucus membrane on the inside. It should be pink. Check the stool. Is it pelleted, soft, or liquid diarrhea? Now put all your signs and symptoms together and either come up with a tentative diagnosis or call your veterinarian with the information.

Figure 6. Observing your herd

By observing that your doe was less than 100% normal you may be able to take action to prevent a serious problem. By checking her out carefully and thoroughly you will be prepared to answer any questions your veterinarian may have if you find it necessary to contact her or him. It's far better to have wasted a few minutes of your time on a healthy doe who was just acting weird than to lose her because of inattention.

Your first awareness of abnormal behavior may be accidental. We've all gone to the barn, done our chores, and then realized that "something wasn't right." In retrospect you realize that the buck's water bucket was still full - and he always drinks over half the bucket. Pay attention when you get "vibes" that something is wrong. Follow through and find the source of the problem. Spend a little extra time each day just observing your herd. Sit with them for a quiet "study" time. It's fun and relaxing and you may notice something important.

NORMAL VALUES

Normal Values for Goats	
Rectal temperature	102.5-104°F (39.2-40°C)
Pulse	60-80/min.
Respiration	15-30/min.
Rumination	1-4/min.
Puberty	4-12 months
Estrus cycle	18-23 days
Estrum	12-36 hours
Gestation	145-153 days

A goat's normal *temperature* will very between 102.5° and 104°F (39.2-40°C). An individual animal may tend to have a normal temperature at one end or the other of the listed range, but for general purposes, any temperature readings within this range can be considered normal. Be sure to shake the thermometer down immediately before using it. A goat's temperature must be taken rectally. Leave the thermometer in place for 2-3 minutes. Expect the temperature to be at the high end of normal if it is taken in the heat of the day or if you had to chase the goat to catch it. A subnormal temperature may also be a sign of disease or it may mean that the thermometer was in the middle of a bunch of pellets and not in contact with the wall of the rectum.

Pulse and *respiration* can vary greatly depending on the goats nervous state at the time the rates are taken. The respiratory rate can be easily taken by watching the rib cage movement and counting the number of breaths per minute. A highly excited animal will have a higher rate than "normal," so use a little common sense and allow for variations with circumstances.

A pulse rate is more difficult to take. It is easiest to take heart rate (which should be the same as the pulse rate) with a stethoscope and count the number of beats per minute. You can also feel for heart beats by holding you fingers tightly over the area of the heart near the bottom and front of the chest. With a little practice, you can take an actual pulse rate by feeling for the pulse on the inside of the rear leg up near the groin, but you would probably need someone to show you just where to feel the first time. Again, this rate can vary greatly with the goats nervous state.

Feel for *rumination* by placing your hand high in the left flank and pressing gently but firmly. As the rumen contracts it will cause the flank to push your hand out for several seconds and will then recede in a wavelike motion. Each contraction should be strong and easily palpable. If you have a stethoscope, you can listen for rumination by placing the head of the stethoscope in the left flank. Each contraction will sound like an ocean wave breaking. Minor gurgles and rumbles don't count. A goat who has eaten recently will have 3-4 rumen contractions per minute. One per minute is normal for animals which have not eaten in the last few hours.

Figure 8. Feeling for rumination

Puberty is defined as that age at which an animal becomes sexually mature. Remember that sexually mature, physically mature, and mentally mature are three entirely different things. Sexually mature means that a doe is capable of becoming pregnant and a buck is capable of impregnating a doe. The listed normal range for when a goat reaches puberty is 4-12 months. But always bear in mind that goats don't read books or lists of normal values. Does have become pregnant at 2 months of age. And it may have been her litter brother that got her in that condition. The key thing to learn from this "normal value" is not to believe all printed lists and to separate buck kids from doe kids and doe kids from adult bucks by 2 months of age.

A female goat is "in season" (in *estrum*) every 18-23 days. This interval between "seasons" is the length of the *estrous cycle*. A doe is in estrum for 12-36 hours and is receptive to a buck for breeding during this time. She ovulates 24-36 hours after the beginning of estrum. Some does have very quiet seasons and you may not know she is in season if you don't have a buck close by to inform you.

Gestation, or pregnancy, lasts 145-153 days and is counted from the day of breeding to the day the kids are delivered. Kids born at 145-153 days are usually quite healthy and ready for life. Kids born at 139-144 days of gestation will be immature and their chance of survival is diminished. Those kids born prior to 139 days are not likely to survive beyond a few hours.

Temperament and Training

A frequent query from the uninitiated is "What is the temperament of Pygmy Goats? Are they like dogs?" There is no simple answer, but they certainly are not like dogs. They do not come when called or fawn at your feet or even "mind" at all! Goats are unique. They are playful, affectionate, endearing, and naughty. Probably "capricious" best describes them. They do as *they* please and not always what you would wish.

Goats are herd animals and need company, preferably another goat. There are situations where a goat will live happily with another animal - horse, sheep, or even a dog - but their preference is a friend of their own species. A friend who thinks as they do and will play silly games with them. There is nothing more entrancing than young kids playing together at butting games or king of the mountain, or just curled up sleeping together. A single goat is a lonely, noisy animal and not much fun for themselves or their owners. Often, new owners can't afford two does to start with and a good alternative is a wether (castrated male) which usually is not as expensive as a doe. Under no circumstances should you buy a young doe and young buck as a pair. Both sexes are very precocious and breeding can take place as early as 2 months of age.

If you already have a small or large herd, you are probably well aware of the "pecking order" that the goats will establish for themselves. Introducing a new goat to the herd will take some thought and planning on your part. Moving to a new home can be traumatic enough without having to fight a large group of bossy goats as they all crowd around to check out the newcomer. It is far kinder to introduce her slowly. Try to choose a fairly mild-mannered goat from your herd and house her separately with the new goat for a few days, until all is peaceful between them. Then when they are all put out together the new goat will at least have a buddy of sorts. Also, it is best to put them all together when the weather is good and they can be outdoors where there is plenty of room, rather than in a small area where the herd can really gang up on her. It will take time for her to become thoroughly accepted, but it will happen eventually. If the new goat is a youngster or a baby, arrange a creep area where she can get away from everyone else and where she feels safe enough to eat her meals and take a nap. The pecking order in a herd is strong, and those on the top of the heap will pick on those more lowly, who, in turn, will pester any new arrival. Try to make things as easy as possible for the new goat.

Pygmy goats are geniuses at undoing locks and gates, escaping from enclosures, and squeezing into areas where you don't want them to go. On the other hand, if you are cleverer than they are and can anticipate (and prevent) some of their escapades, they will settle in contentedly and be your good friends. Patience is the key, along with gentle firmness. They love routine and quickly learn to do as you want if they are always fed at the same times of day and in the same stall. Make any small change and chaos will follow! Work with them quietly and treat them with kindness and you will usually get good results.

These little goats are smart and, with affection and gentle handling, can be trained for many activities. They can be taught to follow with a collar and lead or fitted with a harness for pulling a cart. Many owners enjoy long, woodsy walks with their goats. Others take them for rides in the family car. Some pygmy enthusiasts have kept them in the house and have trained them to use a kitty litter box inside or a dog door to gain access to the outdoors. One drawback, however, is that they love to jump and will sometimes end up on the dining table!

If you plan to show your goats, it is helpful to get them used to wearing a collar and to lead and stand quietly. They should also be unafraid of being handled and examined as a judge would do in the show ring. This can be accomplished by running your hands over them frequently - along the back, under the belly, and down the legs. They will quickly become accustomed to this routine and will be more at ease when shown.

There are a few absolute No No's. Whether the goats are horned or are disbudded, never play butting games with them. It may be cute when they are little kids, to push on the top of their heads and feel their responding push back, but the game will get rougher and more serious (even dangerous) as they mature. Instead, give them attention by scratching the side of their necks or even brushing them. They like to be groomed and that activity has then added bonus of keeping their coats clean and glossy.

Sometimes the goats will be annoying, particularly bucks during breeding season when they become completely mindless for at least three months. Try not to lose your temper. It will not do any good at all. Don't be too friendly with the "boys" when they are acting up. Just do the basic feeding, etc. and stay out of their way. They will return to being nice normal gentlemen eventually!

Most important, enjoy your Pygmies for what they are: fun-loving, responsive little animals who adore their owners.

MANAGEMENT

Housing

Housing requirements vary depending on which part of the country you live in and what the climactic conditions are. In the south where it is warm, a three-sided shed for protection from wind and rain would be adequate, whereas in the northern parts of the country, the goats will need a good draft-free shelter. Pygmy Goats can tolerate very cold temperatures but need protection from rain and snow. A small barn or box stalls in a larger barn work well for the colder climates. An 8' by 10' area can accommodate three or four adult goats. The more space available, however, the better. There are times during a harsh winter spell when the goats will be kept inside for longer periods of time, so consideration should be given to providing enough room to move around freely. If the barn is to be closed up during the winter, there should always be adequate ventilation without drafts. Windows up high above the stalls work well.

Clay or dirt floors in the goat areas are ideal, but wood or cement floors can be managed if they are kept clean and are covered with enough deep bedding in cold weather. Usually the goats will pick one spot where they will urinate and those places should be cleaned out frequently so there is no buildup of urine odor. There is an easy way to check the cleanliness of a stall. Just get down to "goat level" and test the air there! Goats are great hay-wasters and the hay that has fallen out of their mangers makes perfect bedding, so there is no need to buy expensive wood shavings or chips.

Because goats are fastidious eaters and will not eat hay if it has been trampled or soiled, hay mangers are a must. They can be simple wooden structures built into corners or along the side of a stall. They should be at least 15" above the stall floor and the slats 4" to 5" apart. There will still be a certain amount of waste, but saving a little hay with narrower openings won't be worth anything if heads can get caught and terrible accidents happen. More elaborate "key hole" feeders can also be used for both hay and grain. The drawback to feeding grain in a permanent fixture is that it cannot be occasionally scrubbed out as it should be. Pans or individual dishes for feeding grain are more satisfactory, making cleaning easier and also making it possible to monitor how much grain each goat is getting.

Figure 10. Hay manger

Plastic buckets, available at most hardware stores, are the best method of providing fresh water for the goats. The smooth surface is easy to clean. Use the larger buckets for adult goats and the small "paint pails" for young kids. A pig lixit is a great way to provide a constant, clean water source for goats in all parts of the country that don't have to worry about pipes freezing.

A nice addition to their stall, if possible, is a bench for them to play and sleep on. It should be about 18" from the floor and built of sturdy lumber. The goats will love it. And when kids arrive, they will appreciate the safe haven under the bench for their naps.

Figure 11. Sleeping benches

There should be a fenced yard or pasture adjacent to the barn so that the door can be left open in the daytime and the goats can come and go as they please. Sometimes a snow or rain storm will come up quickly and if you happen to be away from home, there is no need to worry. They will head for the barn at top speed.

Dutch doors to the outside are great, giving flexibility for varying weather conditions. In the summer, if the goats are shut in at night, the top of the door can be left open for optimum air. In winter, the bottom part only can be opened in the daytime so the goats have access to their outside area, with the top left shut to cut out some of the cold winds. You can even hang a blanket or burlap from the bottom of the top door to shut out more of the inclement weather. Your goats will have to learn to use this "goat door."

Tethering or staking out goats is a poor idea and can lead to disasters. A tethered goat is a prime target for stray dogs, or the goat can become tangled in the chain or rope. The fencing itself should be the best you can afford. It is better to have a smaller yard with good fencing than a larger area with inadequate fencing. The fence should be at least four feet high (five feet is better) and made of heavy woven wire, stretched tightly on the fence posts. A "horse" fencing with 2" by 4" openings is the best. It is very sturdy and the openings are small enough so that young kids cannot get their heads caught. A fence board at the bottom will prevent dogs from pushing or digging under the wire.

Feeding

As with housing, which varies in different climactic conditions and parts of the country, feeding will very accordingly. In some areas goats have access to year-round pasture; in others, rich alfalfa hay is available; and still other areas are best suited to "dry lot" management. In any case, there are three basic feeding requirements for Pygmy goats: water, roughage, and grain.

Water should be available to them at all times and should be given to them fresh twice a day. They like cool water in summer and, if possible, warm in winter. A mineral salt lick and loose mineral salt are also necessary and will encourage them to drink plenty of water. This is particularly important with bucks and wethers.

Roughage in the form of hay and/or pasture makes up the largest part of their diet, and hay should be fed "free choice." Pygmies are choosey about their hay and prefer it fresh, so chances are they won't eat what hay remains in the manger at the end of the day unless it is fluffed up and new hay added. Find the best quality hay that you can. It should smell fresh and be free of dust. If there are any parts that are the least bit moldy, throw them out. It isn't worth the risk of illness from bad hay to save a few pennies on a bale.

Probably the biggest mistake many new goat owners make is overfeeding grain. Grain is a very small part of their diet and should be fed with care. A 14% or

16% dairy ration (sweet feed) is most commonly used for mature does and growing kids. There are rations made especially for goats and these are fine, too. The bucks and wethers can also have dairy ration in small amounts, but a pelleted form of feed containing ammonium chloride to help prevent urinary calculi is the better choice for all males. Blue Seal offers a "lamb finisher" with ammonium chloride that is designed as a complete ration. If you cannot find Blue Seal products in your area, ask your grain dealer about other brands of feed that will provide this important ingredient.

Dry does and growing kids get 1 cup of grain each at morning and evening feedings and should consume it eagerly and quickly in about fifteen minutes. Lack of appetite may mean either a doe is in heat or the onset of illness, so be sure to pay attention to their eating habits. Do not leave uneaten grain in pens until the next meal. It will just get spilled and may encourage the goats to eat grain from the soiled bedding. Not a good idea!

The amounts for pregnant does should be increased slowly during the last six weeks of their pregnancy up to about 2 cups twice a day. After kidding, the lactating does can be fed the same amount (2 cups) until weaning time and then grain is sharply reduced, or eliminated entirely, for a few days until they begin to dry up.

If you are lucky enough to have good alfalfa hay or rich pasture for your goats, the grain quantities can be reduced, probably by half.

Bucks and wethers should also receive 1 cup of grain twice a day, preferably a ration with ammonium chloride. And they require the same as the does regarding hay, water, and salt.

Whatever your feeding management, it is essential that you know your goats. Each is individual and their requirements may differ. The best test is to *feel* your goats. Looking is not enough. Run your hands over their shoulders and down the back and hips. You should be able to feel the ribs and hip bones but they should be well covered with a layer of flesh. A large belly tells you nothing in regards to their general condition. They should feel slick and smooth.

There are additional "goodies" you can give your goats if you are so inclined. Some people have found that 1 teaspoon of flax seed added to grain at each meal helps maintain a healthy, shiny hair coat; others give their goats vitamins - Clovite is one that is proven; but whatever you decide to give them, be sure the quantities are small and not overdone. Too much can be toxic. And remember that good quality hay, a sensible daily routine, and lots of exercise are the most important of all.

Vaccination

All good goat veterinary management should begin with prevention of disease. Unfortunately, there are many diseases that cannot be prevented by either vaccination or good management, so it is all the more foolish to risk loosing an animal to a preventable disease.

But before you rush out to your barn armed with syringes and vaccines, prepare yourself more completely. A very occasional goat, sheep, or any other animal, will react to a drug. This reaction may be mild or severe and could lead to death within 5-10 minutes of administration of the drug. Your goat may start salivating excessively, exhibit some difficulty breathing, and collapse. Without treatment the goat may die. Buy a bottle or an ampule of epinephrine and tape a small syringe to it. Keep this at hand whenever vaccinating or treating animals. If a severe reaction should occur, you won't have time to hunt for the bottle, much less go out to buy it.

Epinephrine (Adrenalin)
   usually supplied in 1:1000 strength
   dose: kids: 0.1-0.25 cc IM (intramuscular)
            adults: 0.5-1 cc IM

TETANUS

Tetanus protection is available as an antitoxin and a

toxoid. It is important to remember the difference between the two products and to use the correct one at the correct time.

Antitoxin is used to provide temporary protection for 7-14 days. It usually comes in doses of 1500 IU (international units) and can be given subcutaneously. A newborn kid born to an unvaccinated doe may be given 500 IU. All kids being disbudded or castrated should receive 500 IU. If you have an older animal with an injury and don't know its vaccination history, give it 1500 IU. Antitoxin is known to cause anaphylactic reactions on occasion. Always have epinephrine available and get the animal protected with toxoid as soon as possible.

Toxoid will stimulate the goat's immune system to develop antibodies to the tetanus antigen. It should be given to all goats over 4 weeks of age. Start all goats with a series of 2 doses given 4-6 weeks apart. The dose will depend on the products you use. Read the label carefully and give the sheep dose or 1/2 the horse dose. This vaccination should be repeated yearly. It should also be boostered with every serious injury and freshening if it had not been done within the previous 6 months.

SELENIUM

Selenium treatment is not a vaccination but a trace mineral and vitamin supplement. It should be administered at regular intervals so it is most easily included in a vaccination schedule. Bo-Se® (Schering) is the most commonly available selenium injectable product. It is given at a rate of 1cc/40 pounds. Six week old kids should receive 1/2cc. Neonates in need of treatment should receive 1/4cc. The product may be given subcutaneously or intramuscularly. It should be administered twice yearly (or three times yearly in selected cases.) Ideal planning would treat a doe a month prior to breeding and again a month prior to freshening. Different locales in the United States have different selenium levels in the soil and feeds. Check with your local veterinarian to see if your area is deficient.

ENTEROTOXEMIA

Enterotoxemia is caused by *Clostridium perfringens* Type C & D. Like tetanus, it is available as an antitoxin and a toxoid. The antitoxin can be difficult to obtain. It should be used only by veterinarians as a treatment for infected animals. Anaphylactic reactions to the antitoxin are not uncommon.

Enterotoxemia toxoid should be administered initially as a series of 2 shots given 4-6 weeks apart. Most products have a dose of 2cc and can be given subcutaneously or intramuscularly. If the dose is given subcutaneously there is a likelihood that the goat will develop a sterile abscess at the vaccination site. This is a reaction to some of the proteins in the vaccine and should not be confused with caseous lymphadenitis. Administration of the vaccine in non-lymph node areas of the body will help eliminate confusion. Giving the drug intramuscularly will avoid the visible reaction. This vaccine should be boostered yearly. Giving the dose one month prior to freshening will help increase the antibody in the does' colostrum and more protection will be given to the kids. This vaccine is probably not needed in animals on dry-lot management. It is especially important in animals on pastures and with easy access to the grain storage room.

VACCINATION SCHEDULE

Birth
    Tetanus Antitoxin
        for kids born to non-immunized does
    Selenium - if needed
2-3 Weeks
    Tetanus Antitoxin
        for disbudding and/or castration
4-6 Weeks
    1st Tetanus Toxoid
    Selenium
    1st Enterotoxemia (if indicated for your herd)
8-10 Weeks
    2nd Tetanus Toxoid
    2nd Enterotoxemia
6 Months
    Selenium
1 Year
    Selenium
    Tetanus Toxoid
    Enterotoxemia
Every 6 Months
    Selenium
Yearly
    Tetanus Toxoid
    Enterotoxemia

PARASITES

INTERNAL PARASITES

Internal parasites are those animals which live inside the host animal at whose expense they obtain nutrition. Many internal parasites do a great deal of damage to their host (your goat) during the course of the parasitic life cycle. Others will peacefully co-exist with the goat and cause relatively few health problems.

There are four major categories of internal parasites which are of primary concern to goat owners.
1. Gastrointestinal nematodes
2. Lungworms
3. Tapeworms
4. Coccidia

Gastrointestinal nematodes (roundworms) are especially common in pastured animals. Goats kept in a dry lot environment with hay fed in mangers will rarely have a clinical problem with these worms. Goats on pastures, however, require very careful management. Roundworms have a direct life cycle. Adult worms in the intestine lay eggs which are passed in the stool. Under ideal weather conditions the eggs develop to an infective stage in 5 days. Under adverse conditions, they may become dormant and survive many months before becoming infective. Other goats ingest the infective eggs as they eat hay or grasses that have been contaminated with stool. Haemonchus, the barber pole worm, has the greatest clinical significance. These worms inhabit the abomasum (fourth stomach compartment) and are blood suckers. Heavily parasitized animals will develop anemia, edema (swelling) of the lower jaw, edema of the ventral abdomen, and may die, especially if stressed. Other gastrointestinal worms are less "vicious" and may only cause decreased feed utilization and some unthriftiness. A dark black, soft stool may be seen.

Lungworms are also nematodes. They inhabit the airways in the lungs. They cause damage to the airways, eventually blocking them and causing collapse of some portions of the lungs. The airway damage causes a cough and occasionally difficulty in breathing. Only rarely will there be a secondary infection causing a fever or illness. One species of lungworm has a direct life cycle. The worms lay their eggs in the airway. When the goat coughs, they are expelled up the airway to be swallowed and passed out in the stool or they may be expelled in nasal secretions. They develop to an infective stage in 6-7 days. Two species of lungworms have indirect life cycles and require an intermediate host for their development. This means that once the eggs are shed by the goat, they must be ingested by a particular snail or slug, develop to their next stage, and then the snail or slug must be eaten by another goat to cause an infection.

Tapeworms (cestodes) are a relatively overrated parasite. They rarely cause clinical problems in goats. Occasionally young animals with heavy infestations will become pot-bellied and/or show some stunted growth. Tapes shed segments into the stool and these are readily visible to the observant herdsman. (The nematode eggs are microscopic and adult worms are not often passed in the stool, so the herdsman may be unaware of an infection.) The most common tapeworm, Moniezia, is a large worm whose segments are especially visible. Tapes have an indirect life cycle also and a mite is the intermediate host. The goat must eat the mite to become infected with tapes.

Coccidia are single celled intestinal parasites, or protozoa. These organisms are very species specific, so goats can only transmit their coccidia infections to other goats. The goats are not contagious to, or infected by, any other animals. Coccidia have a direct life cycle and most mature animals in a herd are carriers and are shedding infectious organisms in their stool. Kids are most often clinically affected and often show a moderate to severe bloody diarrhea. Some recovered animals will have permanent intestinal damage and may never grow or gain weight properly.

The treatment and prevention of intestinal parasites involves proper herd management and strategic deworming. There is no one right design for a parasite control program. You must diagnose the specific problems in your herd and work with your local veterinarian to design the program most appropriate for your goats. Take some representative fecal samples (1 or 2 adult and 1 or 2 kids) to your vet for analysis. By knowing which parasites infect your herd you can decide which anthelmintic drugs to use and the frequency with which to use them. You must be aware, however, that fecal samples are not a perfect indicator of parasitic infestation. A fecal analysis looks for worm eggs or oocysts from coccidia. Young worms do not produce eggs and early infestations of coccidia do not show oocysts in the stool. A positive fecal can be of great assistance to planning your management program. A negative sample must not be taken as evidence that the animal has no internal parasites.

It is important to remember that goats are considered a food animal. If either milk or meat from an animal is to be used for human consumption, proper withdrawal times must be observed on all drugs. Of the drugs listed in the appendix, only thiabendazole (TBZ®) is approved by FDA for use in goats. Morantel tartrate (Rumatel-88®) is approved as a feed premix. Fenbendazole (Panacur®) is in the process of being approved, but withdrawal times have not yet been established. Each of the other drugs is approved for use in other species (cattle, horses, or dogs). By trial and error and interpolation from the other species, dosages for these other functional drugs have been calculated. Withdrawal times and safety of use in pregnant females and breeding males have not been studied in goats. The goat herdsman must be diligent in his effort to maintain the safety of any meat or milk he produces.

Drugs commonly used in the treatment of the various roundworms in goats are ivermectin, fenbendazole, thiabendazole, levamisole, and oxfendazole. Thiabendazole is approved for use in goats and can be used safely in pregnant animals. Some Haemonchus species are resistant to this drug. Levamisole is approved for use in sheep. It is effective against some lungworms and some resistant Haemonchus strains. Ivermectin is a relatively new drug and is quite effective against Haemonchus, other nematodes, and many external parasites. Cestodes are best treated with fenbendazole or praziquantel.

Young kids may become infected with parasites at 2 weeks of age or younger. Watch for a rough hair coat, pot-belly, loss of appetite, or a ravenous appetite. Some heavily parasitized kids will eat dirt.

If your animals are on pasture, you have additional concerns. The goats will continually pollute the pasture with parasite eggs. The longer they graze the same pasture (or the larger the number of goats on the same pasture), the greater their exposure to parasites. Since goats are browsers they prefer to eat bushes and trees rather than grasses. These are less likely to receive fecal contamination, so are far less likely to be involved in transmission of internal parasites. But goats will eat grasses and, if they are kept on pastures with no brush or trees, they will eat only the contaminated grasses. Pasture rotation, if possible, and avoidance of overcrowding are of concern in goat management.

My small herd of goats live in a dry lot situation. The only fresh greenery they ever have to eat are the weeds that try to grow in their pen in the spring. In fifteen years, these animals have never been dewormed. Babies are treated for coccidiosis. At the same time, there are sheep on irrigated pastures just one mile from me that require deworming every month to prevent deaths from Haemonchus. Take the time to figure out the appropriate parasite control program for your herd.

Measuring Paste Dewormer
Very often the paste dewormers we use come with the dosages measured for larger animals such as horses,

cows, etc. When an accurate dosage is needed for smaller goats and young kids it is difficult to determine just how much is being given. Use the following suggestions for precise dosing.

1. Remove plunger from a 3cc syringe.

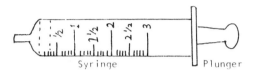

Syringe   Plunger

2. Squeeze dewormer paste into syringe (estimate amount needed).

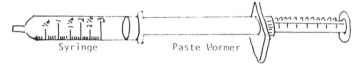

Syringe   Paste Wormer

3. Replace plunger in syringe and push paste down.
4. Check accurate amount. If you have inserted too much paste for dosage, push plunger to right amount and at same time, squeeze extra paste back into original retracted dewormer tube.
5. For really small doses (tiny Pygmy kids) you can be very accurate (1/16 cc) by using a tuberculin syringe for final measure. Squeeze paste into smaller syringe after step 3.

EXTERNAL PARASITES

External parasites are those animals which live outside the host animal at whose expense they obtain nutrition. (Remember that internal parasites, worms, lived inside that same host animal.) Lice, mites, and ticks are the major kinds of external parasites.

Lice are the most common form of external parasitism in goats. Biting lice are small, slightly smaller than the dog flea, and have a red chest and white abdomen. They live close to the skin and will move up along the shafts of the hairs. Split the hair and watch the "dandruff" over a goat's shoulders - if it moves, it's lice. Sucking lice are slightly larger and have a dark blue-black abdomen. They are commonly found on the feet and legs of the goat but may be found on the body. They are easier to see than the biting lice but also tend to occur in lesser numbers. Lice are considered to be species specific except the goat louse which can also infest sheep. Lice cause itching and hair loss. The sucking louse can also cause severe anemia if enough of the parasites are present.

Figure 14. Louse

Mites are much smaller than lice and are most easily seen under a microscope. They tend to burrow into hair shafts and glands of the skin and are more difficult to diagnose than lice. There are four kinds of mites that commonly affect sheep and goats and they all produce different forms of "mange" or "scabies."

Demodectic mange is caused by a species specific mite. Lesions will occur all over the body of the goat. The lesions start as small (3mm) nodules which may increase in size and cause hair loss and a thickening of the skin. There is little or no irritation to the host animal unless a secondary fungal or bacterial infection should occur. Popping one of the nodules and examining the extruded material under the microscope will prove a diagnosis of Demodex.

Sarcoptic mange may spread to man. In animals, lesions tend to start on the head, around the eyes and ears. They may spread to the whole body over the course of 6+ weeks. These mites cause intense itching in both man and goats.

Psoroptic mites are most commonly reported as an ear mite in goats. It may occasionally affect the lower parts of the legs. Severe cases involving the whole body may resemble a fungal infection, but fungal infections rarely cause the itching seen with mange. These cases may require reporting to regulatory authorities.

Chorioptic mange may affect the lower limbs, scrotum, and perineal area. The skin reacts to the mite and oozes a yellow, greasy material. Male fertility may decrease because of the heat of the reaction in the skin raising the temperature around the testes.

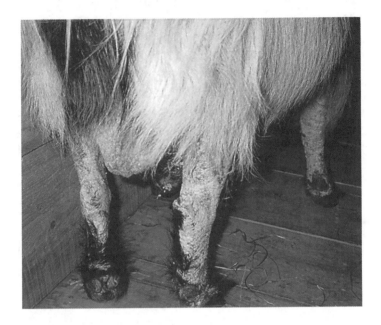

Figure 15. Mange on rear legs

Ticks are blood-sucking parasites. They tend to favor a particular species as a host, but are not species specific. They cause problems for their host in three major ways. A heavy infestation will cause anemia. Some ticks may cause paralysis. Some ticks are vectors for other diseases such as Rocky Mountain Spotted Fever.

Lice, mites, and ticks all are more serious problems in the winter when the host animal has a thick undercoat and longer hair. Many infestations clear up "naturally" in the spring as the animals shed their winter coats. A large part of the population of external parasites and their eggs are removed with the lost hair. The remaining parasites are more easily reached by other treatments. Many insecticides are available in dip or spray form for livestock (Co-Ral® by Miles, Permectrin® by Bioceutic, Expar® by Cooper, etc.). Animals should be treated every 2 weeks for at least 2-3 treatments to break the life cycle of the parasites. Ivermectin is effective against many of these external parasites. It is effective against sucking lice, but not against biting lice. Ectrin, a pour-on, is effective against biting lice. Other management practices, such as rotating pastures, spring cleaning of the bedding in the barn, etc. will also help deplete the environment of the parasites. The best herd managers will plan their spring cleaning and dipping in close time sequence so as to eliminate the greatest number of external parasites. They'll also plan their spring worming at the same time to get the double effect if they are using ivermectin.

Bot Flies are another, often overlooked, external parasite. The bot fly resembles a honey bee. It lays a live larva near the nostril. Goats will become quite annoyed and worried about the flies buzzing around their faces and will stamp their feet and hide their heads and will not eat properly. Once laid, the larva migrates up into the nasal cavity and develops for 2-10 months. The various larval stages in the nasal cavity will cause sneezing and some sinus infections. Up to 75 larvae may be found in each animal. There is no legal treatment for bot flies at this time. Spotton® (fenthion) can be used at 13 mg/lb. I prefer to use a screwworm spray, attach the long red tube from a WD-40 can, cut the red tube to 1 inch, and get one good spray up each nostril. This treatment should not be used on nursing does since it will temporarily damage their sense of smell (which they rely on for recognition of their kids).

Hoof Trimming

Feet are an often ignored and easily forgotten part of the anatomy of the goat. Neglecting proper foot care can lead to lameness and possibly permanent crippling - and a lame goat is of no use to itself or its owner. Goats feet should be trimmed at least every 6 weeks. Some animals with very fast growing hooves or animals that need corrective trimming should be trimmed every 2 to 3 weeks.

A goat's foot should be rhomboid in shape when viewed from the side. The coronary band, where the skin joins the top of the hoof, should be parallel to the ground. The bottom of the foot should be flat with the wall, sole and heel all touching the ground and bearing weight.

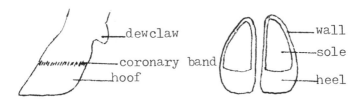

Figure 17. Parts of the hoof

Many tools are used to trim goat's feet - hoof knives, hoof nippers, pruning shears, foot rot shears, etc. Pick the instrument that is most comfortable for you to use, and keep it sharp. Dull hoof trimmers make the job uncomfortable on the goat and difficult for the person doing the trimming.

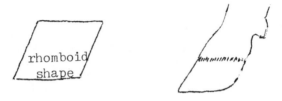

Figure 18. Hoof shape

Start your hoof trim by securing your goat. Either have an assistant hold its head, tie it securely to a fence post (preferably using a halter rather than a collar so the goat won't choke itself), or place it in a milking stanchion. It's impossible to trim feet properly if those feet are racing around a pen. Once the goat is restrained, pick up a foot and hold it in a comfortably flexed position so you have a good view of the bottom of the foot. With rear feet it is often easiest to straddle the goat and then flex the leg and pull the leg up and behind the goat. Clean out any dirt that is wedged in between the overgrown wall and the sole. Next trim the walls of the hoof until they are level with the sole. You may need to trim some extra off the tip of the toe to improve the shape of the foot. Check also between the two hooves near the heels; an overgrown hoof, or goats with certain hoof conformation defects may tend to grow extra hoof into this space. It will cause splaying of the hooves and can become quite uncomfortable for the goat. Trim this excess growth until it is even with the normal inside wall of the hoof. Sometimes the heel will also need trimming to bring it level with the rest of the hoof.

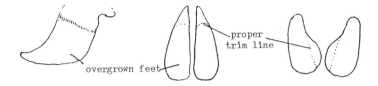

Figure 19. Hoof trimming

Don't be afraid to cut away all excess hoof wall. Often the toe is left too long. As long as you are not cutting below the level of the sole, you needn't worry about causing any bleeding or any pain to your goat. If you need to trim some of the heel, cut only a thin layer at a time and quit if you begin to see a pink color below the surface. If you do cut a bit too deep and cause some bleeding, all is not lost. Put some iodine on the wound and make sure the goat's tetanus toxoid booster is up to date. Only in the case of a very deep cut do you need to consider bandaging or antibiotics.

By including foot care as a regular part of your herd management program, you will keep your goats' feet healthy and make your job easier, too. If feet are trimmed often, it takes only a few minutes to do the job. But once the hooves have overgrown, your job is much harder and takes longer. Try to plan your trimming days to avoid working on does in their last month of pregnancy. They are already awkward enough without having to stand on three legs. And if they fight and kick too strenuously, there is a possibility of damage to the doe and/or her babies. And don't forget that the young kids in the herd have hooves, too. These are ready for trimming by about 6 weeks of age.

Reproduction

HORMONAL CONTROL OF ESTRUS CYCLE

If you are intending to breed your pygmy goat, it behooves you to understand a little about their reproductive cycling. Goats have an estrous cycle with the terms "in season," "in heat," and "in estrus" all being synonymous. The average is 18-23 days from one estrus to the next.

The estrous cycle is regulated by hormones. The normal pattern shows a follicle developing on the ovary and producing estrogen. Follicle-stimulating hormone (FSH) and luteinizing hormone (LH) are produced by the pituitary gland. These two hormones help mature the follicle and cause it to rupture. A corpus luteum (c.l.) develops in the bed of the ruptured follicle and produces progesterone which prepares the uterus for pregnancy. If conception has not occurred, the uterus will produce a prostaglandin which causes regression of the c.l. and will then allow the next cycle to take place. If conception has occurred, the prostaglandin will not be produced and the c.l. will persist and help maintain the pregnancy. Goats require the c.l. for the entire length of their pregnancy.

Occasionally the planned hormonal control will malfunction. A common occurrence is the "five-day heat." A doe will come into season and stand for the buck normally. We carefully enter the dates on the calendar and start planning for our babies in 5 months. But 5 days later, the doe is back in season and standing

59

once again for the buck. He's happy, and so is she, but what does this do to our 5 month plans? Apparently the follicle produced with the first heat did not rupture (ovulate). But since it was ripe, it took only a few days for the hormones to build back up and prepare for ovulation once again. The majority of does *will* ovulate at this five-day heat and you should adjust your calendar accordingly. There is almost no chance that she would have ovulated at the first heat.

This five-day heat can have several different effects on your herd, depending on your particular type of herd management. If you run you does with the buck for breeding and they stay with him for more than the day of their standing heat, you may be assured that he will notice the five-day heat and see that she is inseminated again. If he is the only one who noticed the second cycle, you may just think that she is 5 days late with her delivery. If you hand-breed your goats or only leave the doe with the buck for one or two appropriate days, then it is imperative that you watch carefully for the five-day heat. If you think your does are having 22-28 day cycles, but they aren't conceiving, then you are probably missing that extra cycle. Pay attention; this is one of the few reproductive problems that is easily solved.

Pygmy does usually cycle all year long. Some does may not appear to cycle at all. Check with other breeders in your area to see that their animals are cycling normally. Some does only cycle for 6-8 months of the year - they take a spring vacation. Other does will have "silent" seasons unless a buck is within smelling distance. His odor may be required for her season to be observed and is occasionally required to bring her into season.

CARE OF THE DOE

Does of all ages require good daily care, and the observant herdsman will soon become aware of their needs and idiosyncrasies. Feedings should be at the same times each day so they become settled into an easy routine. They are naturally clean animals and appreciate fresh bedding. Does seldom, if ever, need a bath. Instead, brushing with a stiff brush will keep their coats sleek and shining. Hooves will need to be trimmed about every 2 months (and don't forget to trim the dewclaws at the same time). And, of course, they should be routinely wormed and vaccinated as suggested by your veterinarian. A healthy Pygmy is a happy Pygmy with bright eyes and alert expression and is always delighted to see you.

As noted earlier in the section on Temperament, young does are capable of becoming pregnant as early as 2 months of age. Great care must be taken so that this will not happen. Separate all baby bucks from baby does (even sisters) at 8 to 10 weeks of age.

The optimal age to breed young does is between 14 and 18 months, depending on size and general maturity. Some does grow quickly and could be bred at one year; others are slower to develop and need additional time to attain adequate size. Does bred early are more apt to have a large single kid at their first kidding, making delivery more difficult. By waiting a few more months, the chances are good that the doe will produce twins, smaller in size, and kidding will be much easier.

Size of the doe, however, is not the only criteria to consider. Often a young doe is not emotionally mature enough to cope with the delivery and caring for her offspring afterwards. She may back off, seeming frightened of the wriggling, noisy baby and refuse to let it nurse. It is far better to wait those extra months when the maternal instincts are stronger, than to risk a bad delivery and subsequent "bottle baby."

Pygmy does, generally, are excellent mothers and carefully nurture their offspring if their management has been thoughtfully executed. Mature does should not go into breeding season in fat condition. They should be in good flesh (but not overweight), well groomed, and with all worming and vaccinations up to date.

The first signs of heat are tail wagging and vocalizing. This will usually continue for about 2 days and then, if not bred, repeat itself in about 18 to 21 days. If the doe is living with the buck, he will breed her immediately and if she conceives there will be no further signs of heat.

Does living separately from the buck should be introduced to him at first signs of heat and can either be left with him for a few days or put with him for a breeding 2 times a day for that 2 day period. There are advantages to this method of handling breeding in that you will know exactly when the doe was bred and will, therefore, be able to pinpoint the time of delivery more accurately. The downside is if you have a doe who is difficult to get bred or who doesn't show clear signs of heat, you may waste several months trying to get her to conceive.

Mark down on your barn calendar when you see signs of does coming into heat. You will then be aware of how often and when they are cycling and you can plan just when you want to breed them. This is particularly helpful if you do not own a buck and have to make plans to take your does to a buck at another farm.

Check to make sure your fences are secure so that your buck stays where you want him, and/or your does don't escape for an earlier-than-planned breeding.

Once the doe becomes pregnant, she will require the same daily care as previously, only increasing the grain gradually during the last two months of gestation. As kidding time approaches, prepare a clean stall for her where she can be by herself for the delivery and a week or so afterwards. This will give her time to relax with her kids and for the babies to "mother up" without the threat of other does pushing them around. This is an important time for mother and babies, and it should be as peaceful as possible.

Right after delivery and the necessary cleaning up (which the new mother should do herself) she will appreciate a drink of warm water to replace the fluids she has lost during kidding. A cup of molasses added to the water will give a needed boost of energy. Make sure the doe's teats are not plugged by expressing some milk, then sit back and wait for the kids to nurse. Sometimes the babies will take a frustratingly long time to find the teats, but don't leave until they do. A few more minutes of lost sleep are worth it.

Some does will not be very hungry for 1 or 2 days after they deliver, but their appetite should pick up soon and then it is time to slowly increase their grain to about 2 cups per feeding all during lactation. They require a lot of energy to make the milk their babies consume, so quality hay and plenty of grain are important. (This is another good reason to separate nursing does from the rest of the herd - so you are sure they are getting their fair share.)

When the kids are about 8 to 10 weeks old, are eating grain and hay well, and have learned to drink from the water bucket (make sure there is a small pail available to them), it is time for them to be weaned. Remove them from their dam and put them in a pen where she can't see them. Unfortunately, she will be able to hear them and you will too! At this point, cut the doe's grain out entirely for 2 or 3 days, then offer her a small amount until the udder begins to soften and dry up. Don't try to milk her to ease the pressure. This only encourages her to produce more milk. Most does will dry up easily and without any problems. This process may take 2-4 weeks.

If you are planning to keep any of her kids, you will probably have to keep them separated for 1 to 2 months. The mothering instincts are very strong, and even if she is completely dried up, the kid may go back to nursing and milk will return to the udder. Then you have to start all over again. Better wait a little longer in the first place. Once the kids are completely weaned, they can return to the herd, and most likely dam and daughter will pal around together constantly. Family ties are very enduring, and often an older doe can be found surrounded by her offspring of varying ages - sleeping, playing, or ganging up on an "outsider."

CARE OF THE BUCK

A Pygmy buck is a very significant part of the herd and his management and care require thoughtful planning. Some breeders leave the buck in with mature does year round. Others introduce the buck to the does only during the fall breeding season, keeping them separate from the rest of the herd with another buck or wether for companionship.

Whatever you decide is the best arrangement for housing the buck, there are no alternatives for their general care and feeding. Often, because some bucks are housed apart from the young kids and does, they are given only minimal attention. Then, at breeding time, they are expected to perform with gusto. That may not be possible if they are not in top condition. Proper daily care is essential. They require the same shots, dewormings, hoof trimming, and quality feed as the does and kids. An occasional "hands on" examination can be helpful. Starting with his head, run your hands over him completely - noting if he is in good flesh, coat and skin healthy and clean, and hooves trimmed. He should have increased grain, good quality hay, and fresh water at all times. To ensure that he is drinking plenty of water, have loose salt as well as a salt lick available to him.

In the fall, when breeding is in full swing, the bucks will sometimes "go off their feed." A bit of coaxing or even an extra treat, such as small amounts of coarse ground corn, will stimulate their appetite. In any case, they will eventually eat when they get hungry enough.

One of their least attractive habits is urinating on their muzzles and front legs. This can cause problems with "urine scald" and eventual loss of hair. Try to keep those areas clean by rinsing with warm water, drying well, and then applying a coating of vaseline which will help repel the urine. Make sure he is up to date on all vaccinations and wormings. Your buck is important. Take good care of him.

Puberty vs. Fertility

Isn't it fun to watch young bucklings relating to other goats in their pen? Many of these young males are born with a macho attitude and an idea of what their job will be later in life. Others may be much slower to realize the subtleties of their future job as a breeding animal. It is important for the herd manager to understand that the differences in rate of maturation do not necessarily affect the potential productivity of a breeding male.

Even more important for the herd manager to realize is that puberty is a slowly progressive stage of development. A young buck may be quite capable of courting the girls and of physically mounting and breeding those females who will stand. At the same time, he may not have any viable sperm in his ejaculate. Some physically precocious males may spend several months breeding the females before he actually starts to inseminate them. Other young males will be capable of insemination at just a few months of age.

This may leave the herd manager in a difficult position.
1. Don't count on a young 4-5 month old male as the only sire for the season. He may be a late developer.
2. Don't leave young does in pens with young bucks - or "Murphy's Law" says they are the early fertile ones.
3. Don't give up on a breeding male until he is 12-14 months old. Some males take that long to start passing sperm.
4. Do remember that while puberty may precede fertility by several months, they may also occur simultaneously.

CARE OF THE WETHER

Wethers play a significant role in the pygmy goat world. Often thought of as unimportant or superfluous in other breeds, the Pygmy wether can be a nice addition to a herd or two wethers alone become happy family pets. They are good companions to young kids, a single buck, or a mature doe. They are no threat in the pecking order and peacefully co-exist with almost any other animal including sheep and horses.

They do not have the stresses such as bucks have in breeding season or does have during kidding and, therefore, are usually very healthy and long-lived. The most common problem for wethers is urinary calculi (see page 181) and great care should be taken with their feeding. They require very little grain and can exist almost entirely on good quality hay. Water intake should be encouraged and a mineral salt block plus loose mineral salt helps with that. As with the bucks and does, the water should be cool in the summer and warm in winter. Another possible deterrent to urinary calculi is the inclusion of ammonium chloride in their grain ration. Blue Seal puts out a good product called "Lamb Finisher" which contains ammonium chloride and can be their entire grain ration. About ½ to 1 cup twice a day is plenty. If Blue Seal is not available in your area, check with your grain dealer and see if there is a similar product under another brand name.

Wethers, although not shown in the breed classes at sanctioned shows, have their own age classes and are a popular addition to the shows. They are often shown by young 4H members in the fitting and showing classes at summer fairs and agricultural shows.

PREGNANCY DIAGNOSIS

The first clue that a doe is pregnant is that she does not come back into season 18-21 days following breeding. Watch carefully again at 36-42 days. A small number of does will miscarry at 4-5 weeks. If you noticed the first missed cycle and didn't notice the following season, you may go for 5 months before realizing that the doe isn't pregnant.

Progesterone is produced by the ovaries and can be used as an indicator of non-pregnancy by testing at 19-21 days after breeding. Low levels would indicate that the goat is not pregnant and is getting ready to cycle

69

again.

There are machines that use the Doppler principle to listen for fetal heartbeats. They can first be heard at about 28 days of gestation. The Doppler probe must be aimed directly at the fetal heart. This machine may also pick up enlarged uterine vessels and the "winds of pregnancy" which represent the enlarging uterine cavity.

There are small ultrasound machines for pregnancy diagnosis in sheep and swine that will also work in goats. These machines, however, are not perfect. Since they recognize fluid-filled cavities, they could mistakenly register a positive pregnancy with a full bladder. And, they may miss a late stage of pregnancy since the fetus takes up most of the space and there is little fluid space left.

The last method for diagnosing pregnancy is real time ultrasonography. In the hands of an experienced operator, this is a highly reliable method. The number of kids can be counted and abnormalities in uterine structure can be seen. However, the equipment is very expensive and experience takes a lot of practice.

NORMAL DELIVERY

The process of delivering baby goats is called freshening or kidding. A normal freshening is a joy to witness. The herd manager sits back and enjoys nature at its finest. By watching, he is also there to be aware of any problems that arise as soon as they happen.

The induction of labor is actually accomplished by the fetus. The fetal pituitary produces ACTH which stimulates the production of steroids. The steroids cause the release of prostaglandins by the placenta and the process of expulsion of the fetus begins. Exogenous steroids, those given to the goat by injection or pill, may also stimulate labor. And a rise in endogenous steroids, which may occur in a goat in extreme stress, may also lead to an unexpected delivery.

Each doe will have her own behavior pattern for delivery. A first time freshener will begin to make an udder at about 3 months of gestation. It will fill slowly but steadily throughout the last 2 months. Does who have freshened in previous years may not fill their udder with milk until just a week or two prior to delivery. Both first time fresheners and repeat fresheners may have full, tight udders several days before delivery. An occasional doe will not fill her udder until just after delivering her kids. There is one clue from the udder on delivery time. The teats fill with milk about 24 hours prior to delivery. Do NOT express milk from the teat to prove this to yourself. If you express milk, you remove the body's natural plug in the teat canal and increase the risk of bacteria getting into the udder and causing mastitis. You can feel the teat without expressing it and tell whether it is empty or filled.

A doe may begin pawing at the ground and building a "nest" 24-48 hours prior to delivery. Does don't tell time well though and I have seen does start nest building 3 weeks early. The pawing activity and lying down and rising from the hole do become more frequent as delivery becomes imminent.

Most, but not all, does skip the meal just prior to delivery. She will act restless. She may stop walking, stare off into space, and just stand immobile for 30-60 seconds. She is probably having early contractions that help spread the pubic symphysis and/or dilate the cervix. This is stage 1 of labor. She may pass a thick, whitish mucus plug at about 8 hours before delivering.

Once she is fully dilated, stage 2 of labor, fetal expulsion, begins. Most does lie down for this stage. As the expulsion contractions begin, the doe will stiffen out all 4 legs, you can see strong abdominal contractions, and she will often throw her head back and cry out. Within 1/2 hour of beginning these hard contractions, visible progress should be made.

First a "bubble" of amniotic sac will be seen through the lips of the vulva. The bubble becomes larger and may become grapefruit sized. Sometimes this bubble breaks with a rush of fluid (the water breaks..). Sometimes the baby begins to appear inside the bubble.

As long as steady progress is made, do not interfere. There may be a short pause after the head is delivered before the shoulder comes through. Once the shoulders are delivered, the rest of the body usually slides right out.

Figure 22. Amniotic sac being presented

Afterbirth (placenta) should be delivered within 1-5 hours. The doe may lay down and push like she has another kid. Occasionally, the afterbirth may just slither out with no obvious effort on the doe's part.

Presentation

Kids most often are born with the front legs and head coming first. This is called an anterior presentation. If the kid is delivered with the rear legs first, it is called a posterior presentation. This is a normal and acceptable way for kids to come. A breech presentation has the tail and rump coming first with the rear legs pulled up along the body. Some does will manage to deliver a breech presentation, but most will require assistance.

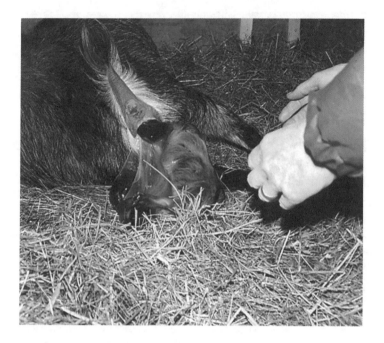

Figure 23. Expulsion of fetus

I DON'T BELIEVE YOU! I GOT A TICKET DRIVING TO THE HOSPITAL, FAINTED WHEN WE ARRIVED, AND NOW YOU TELL ME, "HONEY, I'M _KIDDING_?"

NEONATAL CARE

Pediatric Management

Mother Nature has left the instincts of most Pygmy kids in excellent shape. These kids stand within minutes of delivery and quickly figure out where to find their food supply. But in spite of these remarkable instincts, it is nice for breeders to be present at deliveries and to do what they can to insure success for mother and babies.

Let's assume that your doe is having a nice uncomplicated delivery. Ideally, you will be standing by with a towel to catch the kid so that the freshly broken umbilical cord doesn't get contaminated and dirty. After the doe drops a wet, slippery kid into your towel, you should wipe off its nose and mouth.

75

Sometimes the fetal membranes over the head are quite strong and you will have to exert some effort to tear them off. If the kid is not breathing well, try scratching the sides of its chest with your fingernails. (Cause it enough minor discomfort for it to complain!) Once you are assured that the kid is breathing, make sure that the doe does her share of licking and cleaning the kid. If you do all the cleaning, she will probably assume that you also intend to do all the future work and may reject the kid.

Before the kid starts walking off your clean towel, douse the umbilical cord with 2% iodine. Cover the entire cord and its attachment to the skin on the abdomen. One easy way to do this is to fill a large pill bottle (or baby food jar) with iodine; with the kid in an upright position, drop the cord into the jar and push the jar up to touch the belly; invert the kid and the jar to cover the whole area with iodine. If, for some reason, the cord does not break on its own and the kid is still attached to mother and placenta, you can use some heavy carpet thread or dental floss to tie the cord about 4 inches from the kids belly and again 2 inches past (farther from the kid) that and then cut between your two ties with scissors. If the kid missed your towel and landed in the dirt, or even if it was already cruising around your goat corral with its mother when you first discovered it, iodine the cord anyway. Disinfecting the cord greatly reduces the risk of joint ill. Joint ill, or navel ill, is caused by bacteria entering the kid through the cord and can be a serious or even fatal disease.

Now your kid, with his brown belly, is standing and walking and looking for a drink. Each time he nears his mother, he raises his head and starts searching for a nipple - often starting at the chest and coming out between the rear legs with no success. The doe will talk to him and lick under his tail to encourage him to keep searching. A doe's teats have a natural plug in the opening to protect her from mastitis. At delivery, this plug may be very difficult to expel and occasionally a newborn kid's sucking is not strong enough to open the teat. You should strip both teats and get milk flowing on both sides. This will prove to you that there is milk available for the kid, and it also leaves a milk odor on the teats to help the kid home in on his target. It is important that the kid get colostrum within 1/2 - 1 hour of birth. If you are faced with a weak kid who isn't trying to nurse within that length of time, you would be wise to milk an ounce or so from the doe and bottle feed it to the kid. Often, one small meal is all it takes to perk a kid up. Once a kid has found a teat on his own, you can rest assured that he will find it again whenever he wants.

For the first week or so after delivery, it is a good idea to watch as the kids nurse to see that they are being satisfied. If a kid is constantly attacking the udder and is constantly crying and hungry, it may well be a sign that the doe isn't producing enough milk. Check the udder for edema, mastitis, or just plain lack of milk. If the kids are content and have full bellies, then all should be fine.

Hypothermia is of special concern in neonatal Pygmy Goats. With twins or triplets, the babies will nestle together and keep warm enough except in very inclement weather. Single kids, however, must cuddle up with their dam. If the mother does not sleep with the baby, cold can be a problem even in moderate California weather. Give babies a box with straw to sleep in. Avoid heat lamps if possible.

Figure 25. Kid keeping warm

Checklist after delivery:
1. Iodine cord
2. Colostrum to kid
3. 2 functional teats on doe

Continue to watch udder and health of doe and kids.

Colostrum
Colostrum is the thin, milky fluid produced by the mammary gland during pregnancy and for a few days following delivery. Colostrum is very high in antibodies

and is the only source of passive immune protection for newborn kids. There is no transfer of antibodies from the dam to the kid through the placenta. The kid's intestinal tract is able to absorb the antibodies from the colostrum for approximately 12-72 hours. After that time, the antibodies are digested and utilized as protein. Milking a doe prior to delivery to "see if she has milk yet" is wasting valuable colostrum and is also risking mastitis. Colostrum also acts as a milk laxative to stimulate the intestinal tract and the passage of meconium. Meconium is the dark, mucoid stool passed by the newborn for the first 12-24 hours of life.

If a kid is not able to receive colostrum from his mother (the dam has CAE or died during delivery or has had a mastectomy), there are several options available. Sheep or cow colostrum can be fed. The kid will not receive the same immune benefit of the antibodies, but the rich protein and laxative effect will still be of benefit. There are commercial colostrum products available, but their benefit is questionable. If you have more than one goat or are planning to breed your doe on a regular basis, it would be wise to save a cup or so of colostrum and freeze it for future use. (The colostrum must be collected within an hour or two of delivery while the antibody levels are still quite high. As the doe starts to produce more milk when the kid nurses, she will dilute the antibodies in the colostrum.)

Figure 26. Bottle feeding

Bottle Feeding

Ideally, newborn kids should be left to nurse their dams. This gives them the best possible start in life and the mother will quickly teach them to eat hay and grain and become part of the herd. However, there are occasions when nursing the dam is not possible or desirable, as with a CAE positive doe; or you have a doe producing triplets or even quads, and although she is an excellent mother, she simply cannot supply enough milk for all her babies.

If you are thinking of bottle feeding, plan to start the kid (or kids) right away. It is much easier to teach a kid to accept a bottle before they find the joys of nursing their dam. With multiple babies needing supplemental feedings, they will usually learn to alternate between mother and bottle once they have discovered that the bottle contains delicious milk.

Goat milk is obviously the best choice for feeding kids, however, with the possibility of CAE contaminated milk, you must consider carefully what you will use. Properly pasteurized goat milk is good as is canned goat milk, however, the canned milk is very expensive and the cost adds up over the weeks that the kids will need the milk. Many people use Lamb Replacer, Land O' Lakes Kid Milk Replacer, and even plain homogenized cow milk from the super market - all with some degree of success. *Canned milk must be diluted half with water before feeding to the kid.*

The milk should be heated to approximately 102°F. This temperature is warmer than would be fed to a human baby but not hot enough to burn your wrist when testing it. The nipple should have a hole large enough so that the milk flows fairly freely but not so fast that it chokes the kid.   There a new nipples - Pritchard Teats - now available through many of the farm supply catalogs: Omaha Vaccine Company, KV Vet Supply Co., and Caprine Supply to name a few. These little nipples screw onto a soda bottle and are perfect for Pygmy kids. They are well worth sending for. Use nail polish to mark ounce measurements on the bottle.

For the first few days after birth, the feeding schedule is fairly intensive with many small feedings at short intervals. The babies are so little and their stomachs can't hold much milk at one session (about 1 to 2 ounces), so frequency is the answer here. As they drink more at each feeding, the times can be extended so that fewer and fewer feedings are needed. Most kids need only be fed during daytime hours. Nighttime feedings are only necessary for kids in less than perfect health. Approximate feeding times and amounts are indicated in the following chart:

Age	# feedings	Amount	Times
1 week	5	1-3 oz.	8,11,2,5,8
2-3 weeks	4	4-5 oz.	8,12,4,8
4-5 weeks	3	5-6 oz.	8,2,8
6-7 weeks	2	8-10 oz.	8,8
8 weeks	1	Dilute with water, decrease amount	

Should be completely weaned by 10 weeks.

The biggest mistake with bottle feeding Pygmy kids is overfeeding on milk and continuing the bottle too long. The kids should be encouraged to eat hay and grain as early as possible. If they are full of milk all the time, they won't bother to experiment with other foods. Kids with their mothers learn quickly to eat grain and hay, and if kept in with the rest of the herd, bottle babies will learn more readily, too. A "creep" can be constructed that will allow all kids (including bottle babies) to have an area all their own where grain, hay and water are available all the time without competition from older goats. The kids who have learned to nibble hay and grain from the dams will teach the slower bottle babies to join them. The creep method will also allow the bottle kids to become part of the herd more easily.

The creep can be a very simple gate arrangement with vertical slats about 4.5 inches apart partitioning off an extra stall or the corner of a stall. It should be sturdily built with 2 x 4's as the older goats will try their best to get into the kid's area where the grain is. The creep can be a temporary structure that can be folded up and stored away until needed again.

Bottle feeding kids can be a tricky business, and the key to success is keeping a very watchful eye on their development. These kids do not have the natural immunities that is provided to kids nursing their dams, so they are far more susceptible to parasites of all kinds. This is where feeling your goats is important. Sometimes kids will appear very fat with a large belly, but on closer examination they may be very emaciated with protruding hips and ribs, the large belly being an indication of an infestation of internal parasites. Coccidia is probably the most common problem in baby Pygmies. There are, however, other very damaging internal parasites that should also be considered. Albon for Coccidia and ivermectin paste for general deworming will be your best friends during bottle feeding. Use them with caution but with some frequency if you are seeing (feeling) problems. Eventually, the kids will produce their own immunities and will outgrow the need for the various frequent worming treatments.(See page 143 for more information on coccidiosis.)

A final word on weaning - do it as soon after 8 weeks as possible. The kids should be eating grain and hay, drinking from the water bucket with regularity, and the bottle feedings should have been almost eliminated by

8-9 weeks. Some kids are quicker than others to master these feats, so you will have to be somewhat flexible, but don't make the mistake of thinking that bottle feeding is "cute" and continue it on and on. It isn't cute, and you are not doing them any favors. Pygmy kids need to develop their rumens and their independence for a long and happy life.

Figure 27. Kid learns to eat from Mom

Folded Ears
Occasionally young kids are born with one or both of their ears "folded" or tipped in an unnatural way. These ears need prompt attention before they have had time to become permanently deformed.

Gently fold the ear into its proper position and wrap

scotch tape loosely around it to hold it in position. Leave the tape on for four or five days then remove. By then, the ear should have assumed it correct shape, but if not, tape again for a few more days.

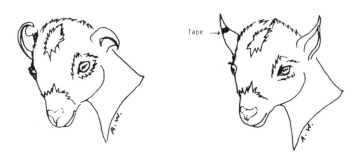

Figure 28. Folded ear and taped ear

Contracted Tendons in Newborns

Not uncommonly, a kid will be born with one or both front legs flexed and unable to completely extend. These babies have spent 5 months in a confined uterus and most likely have had their legs flexed under themselves in a sleeping position. The muscles that bend the front legs are called flexor muscles; those muscles that straighten the front legs are called extensor muscles. Since the flexors have not had to stretch out in the uterus, they may become contracted and shortened. The baby may be physically unable to fully stretch them and extend the legs immediately after it is born. This leaves the leg bent at the knee and the pastern. If the flexor tendon is only slightly contracted and the hoof still rests correctly on the ground, then the problem will resolve itself in a few days to a week. As the baby stands on its hoof it will naturally stretch the flexor tendon back to a normal length. If the contraction is severe, the hoof may turn under and the weight may be placed on the front of the hoof. In this case, nature is not working with the animal and there is no tendency for the tendon to stretch. The herd manager must step in and help.

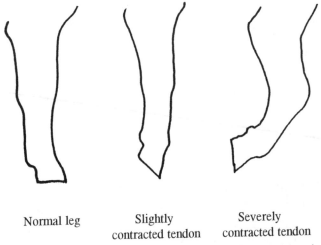

Normal leg   Slightly   Severely
contracted tendon   contracted tendon

Figure 29. Normal and contracted tendons, side view

The object of treatment is to stretch the flexor tendon. This is most easily done by applying a splint to the back of the leg. A splint can be made from PVC pipe (usually 1 inch diameter for Pygmy kids). Measure the normal leg from the bottom of the hoof to the elbow. Cut the same length of pipe. Then cut the pipe in half lengthwise so you have two long troughs. Smooth the edges and pad the trough with cotton. Using tape and gauze, attach the padded pipe to the back of the contracted leg. If you cannot straighten the leg enough for the knee to rest comfortably in the trough, apply your gauze as snugly as possible. In 12-24 hours, you can add more tape over the original and pull the knee completely into the splint. The steady tension applied by the splint will fatigue the flexor muscles and slowly allow them to stretch. The splint can be removed in 24-48 hours. The leg will look weak and wobbly for 6-8 hours while the muscles restrengthen and then the baby will be as good as new. Very rarely the splint will have to be replaced for an additional 24 hours. (I treated one Alpine kid with 3 legs with contracted tendons. The splints were applied within 8 hours of birth and the kid required assistance in getting up and down. He wore the splints for 3 days. He is now 11 years old and I would defy anyone to notice any conformation problems with his legs.)

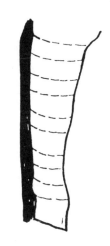

Figure 30. Splint

The question, of course, is why do we see contracted tendons at all? I think many cases are strictly due to uterine position. Other cases respond beautifully to selenium supplementation and are surely due to selenium deficient muscular dystrophy. Heredity does NOT appear to be involved. I suggest treating mild cases with selenium and T.O.T. (tincture of time) and the more severe cases with selenium and a splint. One day with a splint may not be needed, but won't cause any trouble if you have any doubts. If the problem is not corrected very early, it becomes much more difficult to treat. The splints must be left on for longer periods of time and the contraction may not lengthen completely.

Babies with lateral deviations of their front legs (pasterns that bow out away from the body) are best treated with selenium injections. This problem is not usually due to uterine position. Again, a splint won't hurt but it won't solve the problem without the addition of selenium.

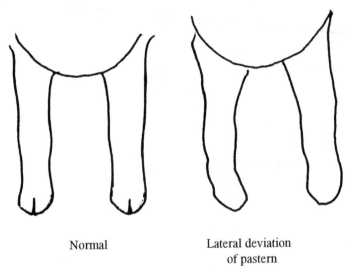

Normal

Lateral deviation
of pastern

Figure 31. Pasterns

The Intersex

For years we have heard that the male of the species determines the sex of the babies. But it's time for the feminists in the group to stand up and take part of the credit too. And if both the doe and the buck are claiming responsibility for the sex of their kids, who claims the credit for the "intersex" kids?

A quick review of our elementary genetics shows that a goat with two "X" sex chromosomes is a doe and a goat with an "X" and a "Y" sex chromosome is a buck. Since each parent gives one sex chromosome to its offspring, it is obvious that only the buck can pass on a "Y" chromosome to produce a male kid. Thus, he takes the credit for sex determination! But the vaginal and uterine environment in the doe may be hostile to "X sperm" and favor the "Y sperm" (or vice versa). And the sperm may have to spend several days waiting for a ripe egg to be produced. The "X" and "Y" sperm also travel at different speeds (the "Y"s are faster) and have different average life spans (the "X"s live longer) and both these factors can be very important on the final outcome of the sex of the kid. Early breeding favors females and late breeding favors males.

But what happens if several eggs are fertilized and the law of averages holds true and two buck and two doe kids develop in the same uterus? Usually, the boys will develop as healthy little bucklings and the girls will develop as healthy little doelings - each happy and content in its own amniotic sack. Occasionally, however, there will be some placental sharing and exchanging of hormones in the developing fetuses and the doe kid may be adversely affected by her brother(s). A female goat kid born twin (or triplet or quad) to a buck may be a freemartin. Her chromosomes say she's a girl, but her body didn't follow through. These freemartins can be subtle or severe. Some does will look like bucks or wethers by being very muscular and fast to grow, but her teats and vulva may stay abnormally small as she matures. She will not come in season and may have a reproductive tract that stops just a couple of inches into the vagina. Her ovaries are non-functional and may actually resemble a testicle. Freemartins are sterile, but their brothers will be normally fertile. Freemartinism is the genetic fault of neither the sire nor the dam. It is strictly the result of the placental attachments in the uterus in a particular pregnancy. No change need be made in a breeding program following the birth of a freemartin.

Hermaphrodites, on the other hand, are a more serious problem. They are extremely rare and have both ovaries and testicles in one animal. Pseudohermaphrodites are more common, especially in polled goats, and have the gonads of one sex with the external genitalia of the opposite sex. It is most common to have a goat appear to be a doe but to have male gonads (testicles). There is an apparent genetic component to the development of hermaphrodites and pseudohermaphrodites, and it would be unwise to repeat the breeding that produced either problem. However, either parent could be bred to many different partners with small probability of producing similar problem offspring.

Intersexual problems are an end in themselves. They will not reproduce. A freemartin should cause *no* change in your breeding practices and a pseudohermaphrodite should only cause a different sire/dam combination to be used in the future. There is no cause for panic and no need for pointing the blame. However, the ethical considerations of selling doe kids for breeding could present some new problems. Most of us sell kids at a very young age and would have no reason to suspect intersexual problems. At the same time, our buyers are expecting breedable does. I, personally, would replace any doe I had sold which turned out to be either a freemartin or hermaphrodite or pseudohermaphrodite. Each breeder needs to consider the problem before it arises.

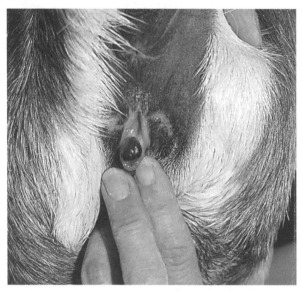

Figure 32. Enlarged clitoris in a hermaphrodite

Disbudding

If you don't want horns on your adult goats, you have the option of disbudding your babies or dehorning your adults. Disbudding is much less stressful to the goat and is the method of choice.

Disbudding is a procedure which destroys the young horn bud and prevents it from growing. A hot iron (disbudding iron) is applied to the horn bud and a circle of tissue is burned around the entire developing horn. The blood supply to the horn bud is destroyed and the horn is unable to grow. Many disbudding irons are available from commercial livestock suppliers.

Since disbudding is destroying the young horn bud, it must be performed on a horn *bud*, not on horns, so it must be done on young kids. Bucklings usually need to be done by 1-2 weeks of age. Does can wait until 2-4 weeks of age. Exact time for the procedure is determined by the individual animals and the size of their horn buds. The diameter of the base of the horn (where is connects to the head) must be smaller than the inside diameter of the disbudding iron. Bucklings' horns grow much faster than a doelings' so brothers and sisters often cannot be disbudded on the same day. Some bucks need disbudding almost at birth (I do wait 2-3 days) and some does can wait until 6-8 weeks of age.

Once the horn buds are large enough to feel easily and not too large for the iron, the kid must be restrained and the hair clipped over and around the horn. The clipped area should be wiped off well with alcohol. The iron should be preheated for at least 5 minutes prior to use. With the kid held very firmly so the head can't move, the iron is applied over the first horn. Push it firmly over the horn with a slight rotating action. Continue the pressure and rotating for 9-10 seconds and then remove the iron. (Never leave the iron on for more than 10 seconds or enough heat may build up to burn the brain inside the skull.) Examine the "moat" you have created around the horn. It should extend through the skin all the way to the skull. The skull will have a dry, dull look to it. If there is still glistening tissue in the depths of the "moat," then there are still blood vessels crossing in to the horn bud. Apply the iron again for 5 seconds at a time. Tip the iron so the edge leans a little in toward the unfinished areas and remember to rotate the iron slightly while it is applied. When the "moat" is completely dull and dry looking, you are ready to do the second horn. You may want to wait 30-60 seconds to allow the iron to reheat slightly before continuing. (**Note:** Individual irons differ in heat production. Some may need to be applied for a shorter length of time than 10 seconds.)

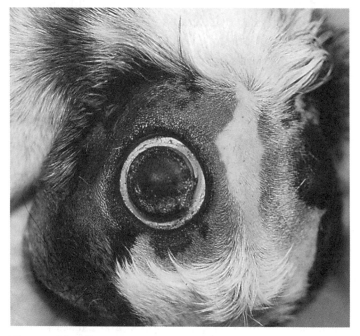

Figure 33. Kid with one side disbudded

Now the job is done, and there is no real aftercare necessary. But there are still a few things for you to do. I like to apply a burn gel to the "moat" as I finish. I assume it make the goat's head feel a little better, and I know it makes me feel like I've helped. Make sure the kid has a tetanus antitoxin shot. Burns do create an environment that the tetanus germs like and I have seen kids with tetanus following disbudding. When you are ready to return the kid to its mother, try to let her sniff the kid's tail area and tuck the kid's head back under her to nurse. You should have no problem with her rejecting the kid. Occasionally, if she sniffs the kid's burned head first, she may try to keep it away.

The entire procedure of disbudding should take less than 10 minutes. It is relatively easy to perform once you master the technique. But there are a few problems that we will discuss before you consider yourself a pro.

Some owners will argue that the kid should be anesthetized to be disbudded. I have tried general anesthesia, local anesthesia, and no anesthesia in my practice and on my own goats. General anesthesia is a serious procedure in itself, especially in young animals, and I don't feel that it should be utilized unnecessarily. The kid has much more trouble recovering from the effects of the anesthesia than from the disbudding. And there is a much greater risk of rejection by the dam when the kid is anesthetized. (She is much more concerned about a baby who is unstable on his feet than one whose head smells a little funny.) Local anesthesia causes pain while it is being applied and causes itching as it goes away. Most kids complain about the restraint for clipping or local anesthesia almost as much as about the actual disbudding. I'm not trying to say that the disbudding procedure doesn't hurt; I am saying that doing a disbud quickly and without any anesthesia is the lesser of the possible evils. As soon as you put the kids back with their mother, they seem content and show no signs of abuse or discomfort.

The other serious problem with disbudding is the scurs that may come back to haunt you. A properly disbudded kid will take 6-8 weeks to heal the burned area and start growing in hair in the shaved areas. By three months you should not even know that anything had ever been done - except there will be no horns. If you disbudding was not performed correctly, or if the horn was a little too big, scurs will develop. Any horn bud tissue that was not killed will try to grow. Male animals tend to have a small, pointy extension on the inside toward the front of the horn. This extension is easy to miss with the disbudding iron. It will grow in weird and unusual shapes, depending on just how much of it is left. Intact bucks are much more likely to develop scurs since the male hormones stimulate horn growth. A wether or a doe can also develop scurs, but they work less hard at it! If you do notice a small scur developing within the first few months after disbudding, you can repeat the disbud procedure (as long as the scur is small enough to fit inside the iron). Some scurs are very loosely attached and can just be broken off with your fingers. Others become so big that they must be removed by dehorning. Avoiding them is the best way to go.

Figure 34. Scurs

Tattooing

Permanent identification of goats can be difficult to achieve. Ear tags get ripped out, branding destroys the animal's esthetic value, and tattoos become unreadable. Electronic identification implants may well be the future of identification, but are not yet commonplace. Tattooing remains the method of choice in goats at this time.

There are no guaranteed ways to permanently tattoo a goat that I know of. Some animals seem to take delight in making a tattoo disappear while others keep theirs legible forever.

Prepare your tattoo instrument first. Use 1/4 or 3/8

inch digits. Clean them with alcohol and arrange then in the correct order in the instrument. Double and triple check that you are using the correct letters and numbers in the correct ear. Make a practice run on a piece of paper just to be certain they read correctly. It's extremely easy to end up with the whole thing reading backwards (2AQ instead of QA2). Either get an assistant to restrain the goat's head or put her in a stanchion. Clean the inside of the ear with alcohol and wipe it dry. Spread some tattoo ink (green seems to be preferable to black in dark eared goats) on the ear where you intend to place the tattoo. Pick a spot away from the hair at the edge of the ear. Apply the tattooer and squeeze. Try not to puncture all the way through the ear, but when you release the instrument the ear will probably be stuck to the prongs of the tattoo digits. Gently peel it off. Don't be upset if there is some bleeding. Apply more tattoo ink and massage the ink into the holes with your thumb. Now repeat the process with the other ear. And don't forget to make sure the goat's tetanus toxoid vaccine is up to date.

If you happen to be an advocate of tail tattooing (which I'm not), you just follow the same procedures as for the ears but use the web at the sides of the tail close to the body. Try not to distort the skin on the tail other than to stretch it out flat when you apply the tattooer.

Iron deficiency
Occasionally you will find your 4-8 week old kids eating dirt. This may be due to parasitism or it may be due to an iron deficiency. Some kids will eat enough dirt to cause moderate to severe intestinal distress. These kids lose their appetite for their regular food but

may still eat dirt when given the opportunity. If iron deficiency is the cause, an oral iron supplement such as Fer-in-sol® will stop the problem in just a few days. With parasitism, administration of dewormers is indicated.

FETAL DEVELOPMENT

Not all pregnancies will end in a full-term normal birth. There is a long list of possible causes for abortions in goats. It can be very difficult to diagnose the cause even with proper laboratory equipment and good specimens. But it can be helpful, and even interesting, to know at what stage of development the fetus died. Did it die at 2 months but was not delivered until 5 months; did death and delivery occur simultaneously; or was there an early delivery of a live kid? These questions can be difficult to answer but can be of great assistance in deciding the cause of the problem. The following is a rough outline of the development of the goat fetus that can be used to "guesstimate" the age when development stopped.

Fetal Development

Month of Gestation	Size	Weight	Comments
1 mo	0.25-0.5 in.	0.1-0.25 oz.	Facial features visible, limb buds growing
1.5	1.0-1.5		Eyelids visible, limbs more slender, neck elongates, cephalic prominence reduced.
2	1.5-2	1	Pigmentation and horn pits present
2.5	2-2.5	4	Hair on eyelids and lips
3	3-4	8-12	Hair on tail and muzzle
3.5	4-5	0.8-1.2 lb	Fine hair on body, hair on legs, nostrils open, tooth buds visible
4	5-6	1.5-2.5	Haircoat complete, incisors erupting
5	5-7	2.5-4	Incisors erupted

MILKING

To milk or not to milk......Our Pygmy does really let us make this choice. Dairy breeds have been bred for milk production and have reached the point where they produce much more milk than is needed to suckle kids. These dairy does must be milked to prevent illness, mastitis, and damage to the udder. On the other hand, Pygmy does produce enough milk to raise fat, sassy kids, but not enough to share much with their human families. However, our Pygmies will continue to produce very rich milk after their kids are weaned if their owner is willing to milk them.

Goats tend to increase in milk production during successive lactations and by her second lactation, a Pygmy goat may easily produce 4 pounds (2 quarts) of milk a day and this milk may contain up to 8-9% butterfat. Goats milk has long been claimed to be more easily digestible than cows milk and is often recommended for babies and some elderly or sickly people. The only problem with milking Pygmies lies in how to get the 2 quarts of milk out of two tiny teats on one small doe.

Milking a goat is an experience that should be enjoyable to both goat and milker. It takes a little bit of skill and a lot of care to do the job properly. A less than ideal milking procedure may lead to mastitis and/or an unhappy goat.

Milking must be done in an organized fashion. A milking stanchion isn't a requirement but is definitely advantageous. The stanchion confines the doe at a level which allows milking without breaking your back and usually provides a small feeder which will hold grain to keep the doe happy while she is being milked. She should NOT be fed all the grain she wants but only enough to help keep her milk production up; usually this amounts to 1/2-1 pound of grain per pound of milk produced. Remember, 1 pint of milk = 1 pound. Excess grain feeding will produce fat and not more milk.

Before milking, the udder should be washed with warm water or any udder wash available for cattle and then patted dry. The first few squirts of milk should be aimed at the ground so that the mucus plug in the teat canal will be removed. Then the udder should be thoroughly emptied into the milk bucket. Once you think you've milked her dry, massage the udder firmly and milk some more. This massaging and milking will help release the milk from all the alveoli in the udder. If the udder is not completely emptied at each milking, mastitis may result. Lastly, the bottom one inch of the teats should be dipped in a teat dip. Many different teat dips are available; some are antiseptic and some put a latex seal on the end of the teat. All dips help protect against mastitis.

One more step in the milking procedure must not be forgotten. Your doe has produced several pounds of delicious milk - which starts to spoil immediately. You should strain your milk through milk filters (available through Sears Farm catalog, feed stores, and dairy supply stores) and you MUST cool your milk quickly. Unpasteurized goats milk which is refrigerated immediately after milking should stay fresh for about 1 week. If you wish to pasteurize your milk, heat it to 145° for 30 minutes. Drink your goats milk; it's delicious and nutritious.

Now you've been milking your doe twice daily for a couple of months. The milk is delicious and you and she are enjoying the quiet time together. But her production is dropping and you feel it is time to quit milking her. Besides, your next doe is ready to wean her kids and if you milk too many of your does, you'll have excess milk and aren't really ready to start making cheese. Or maybe you've just sold your doe's twins and aren't sure what to do about her udder which you know is going to fill to its maximum capacity and you don't want to start regular milking. It's time to dry up these does.

Drying up a doe is quite easy. When you wish to quit milking or are weaning kids, first stop all the doe's grain ration but feed her hay and water as usual. A few days later, you take her kids away or don't show up for her normal milking, and you have dried her up. The udder must fill with milk and put back pressure on the milk secreting glands to give them the message to stop milk production. The doe will be uncomfortable for a few days, but usually within one week she will be comfortable again and the udder will begin shrinking. Do NOT milk a little out to relieve the pressure on her bag, and do NOT milk her every few days. This defeats the process of drying her up and may lead to mastitis.

Necropsies

No matter how careful you are as a herd manager, sometimes an animal will die. The reason for the death may be obvious (such as the pack of dogs standing over the carcass) but more often you will not be aware of any good reason. The decision to have a necropsy done may be difficult - I have a great deal of difficulty practicing what I preach when I lose an animal in my own herd - but it is foolish to lose an animal and not know why.

Necropsies may answer your questions concerning cause of death and those answers may save many other animals in your herd. The most information will be obtained from a necropsy done by your state pathology laboratory or by your veterinarian who may send samples into the state lab. However, just a gross necropsy (looking only at what can been seen with the eye - no samples sent for microscopic or bacteriological examination) will often provide some answers. If expenses are a particular problem, you can even do a gross necropsy yourself. It is best if you have at least seen inside a dead animal before, but don't let the lack of experience deny you a possible answer to a problem now. If you open up the carcass and see something that you are unsure of, call your vet and ask what normal is. Depending on the size or what you are examining, you may even take it to your vet for an opinion. The next time you will know more of what is normal.

VETERINARY
CARE
TECHNIQUES

Signs to Check for Illness

Diseases manifest in many different ways and the herd manager must be aware of the various aspects of the goat to watch so he will find problems before they become as serious. And if a veterinarian must be called, he will be prepared to provide answers to the vet's questions.

First, observe the goat's general attitude. Notice if she is alert and interested in her surroundings. Watch her response to other animals and people around her. Subtle changes in attitude are often the very early warning signs of a disease process.

Next check her appetite. It is important to notice not only how much she is eating but what she eats and how readily. Some diseases will cause a goat to go off grain but she will still eat hay. Others will not eat hay but may still eat some specially tempting browse.

Check her droppings. Diarrhea, constipation, change in size or color of fecal pellets may all be indicative of different problems.

Keep a thermometer available. If you are suspicious of a sick goat, take her temperature. If her temperature is normal, she may still have a problem, but a fever is a definite indication of trouble. However, in the hot part of the day or after a hot and stressful car ride to the veterinary clinic, the temperature may be at the high end of the normal range.

Look at the goat's gums or pull down an eyelid and look at the conjunctiva. Both should be a light pink color. Bright red or pale white are not healthy colors. If you are in doubt about normal color and have another goat available, compare colors.

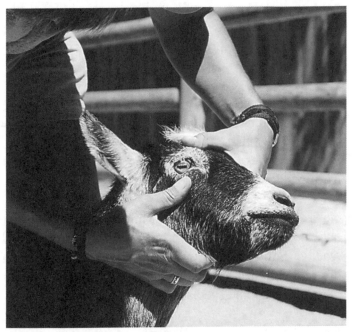

Figure 35 Checking mucus membranes for color

One last parameter to check is rumination. If you see your goat chewing a cud, she is ruminating normally. Do verify to yourself that she is chewing a cud and not just grinding her teeth. Watch also to see if she is eructating (belching).

Restraint

Knowing what needs to be done to your goat (shots, foot trimming, etc.) is only half the battle. Since very few goats are obedience trained, it is highly unlikely that yours will stand quietly while you subject it to any treatment other than brushing. There are several different methods commonly used to restrain goats.

Figure 36. Using the eye hook and platform

1. Collar and "eye" hook - Probably the simplest way to hold your goat for most procedures is to use a very short chain and snap to connect the goat's collar to an eye hook which is attached at the appropriate level to a *firm* post or wall (**Figure 36**). Once attached,

the goat can be squeezed up**Figure 37.** Goat halter

109

against the fence or wall by the person doing the work and can then only move forward or back a few inches. It cannot rear up very far nor lie down. The ingenious herd manager (or the ones with bad backs) can build a platform and install the hook so the goat stands on the platform at a more convenient level for working. A goat halter (**Figure 37**) may be used instead of a collar.

Figure 38. Stanchion

2. Stanchion - If there's a carpenter in the family, the stanchion is an ideal way to restrain most goats (**Figure 38**). There are many varieties of stanchions, but the bottom line is that it is a device that raises the goat 12-18 inches above ground level and holds its head such that it can't move forward or back. Usually the head is held between two bars and can move up and down. Often there is a grain feeder attached so the animal could be fed while restrained in the stanchion. Not all stanchions prevent sideways movement, so may have to be set up near a wall to keep the goat from falling off the side.

3. Leg up - If there are two people working with the goat, a quick way of restraint is to pick up one front leg (**Figure 39**). It is important that the handler pick up the front leg on the opposite side of the goat so he can hold the goat tight to

Figure 39 Holding one leg up

him while holding up the leg. The slight loss of balance should make the goat more tractable and less powerful if she continues to fight.

4. Miscellaneous - There are other less commonly used methods of restraint. An animal can be backed into a corner with the handler straddling the shoulders and holding the head up high. This is great for taking blood samples and examining front legs. Smaller goats can be held on their sides in a prone position for disbudding, bandaging legs, etc (**Figure 40**). The animal is laid down on its side with the handler positioned at its back. He reaches across the goats neck and takes hold of the bottom front leg just below the elbow. By applying leverage up on the leg and down on the neck, the animal's front end is easily restrained. The handler's other hand will usually hold the two hind legs to keep the rear end from twisting under the goat in an attempt to rise. When holding two legs in one hand, it is important to put your index finger between the legs with your thumb around one leg and your other three fingers around the other. This will give you a firmer

grip and will help prevent injury to the legs.

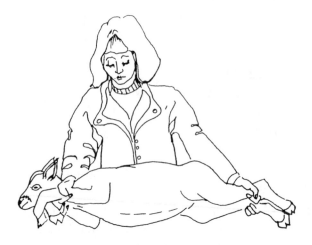

Figure 40. Holding a goat on its side

Medications

In spite of, and occasionally because of, the best management, your goats will require medication. Medication comes in pill, oral liquid, and injectable forms. You should know how to properly administer each of these.

1. **Pills** - A goat's mouth does not open very wide, so you can't just poke a pill down like you would with a dog. Remember that your goats ruminate and chew on hay the better portion of their day. The same rear teeth that macerate hay are capable of making mincemeat of your fingers. There is an instrument called a "balling gun" that can be used to safely administer pills to your goats. The balling gun (Figure **Figure 41**) is a long narrow instrument with a plunger in the center. There is a pocket at one end to hold a pill and the plunger is activated by your finger or thumb at the other end.

Figure 41. Balling gun

The loaded instrument is inserted in the side of the goat's mouth and directed up along the inside of the cheek teeth to the base of the tongue at the back of the mouth (**Figure** 42).

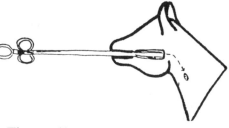

Figure 42. Balling gun in use

Then you push the plunger to eject the pill. If the pill is just past the base of the tongue, the goat will be forced to swallow it. If it is placed too far forward on

113

the tongue, the goat may chew it up and spit it out.

One concern of goat herd managers has been the possibility of ejecting the pill into the trachea. The anatomy of the goat's throat would make this very difficult to do, but if you keep the end of the balling gun aimed toward the side of the throat rather than right down the center, it will be almost impossible for the pill to go anywhere but down the esophagus.

Another option is to mash up the pill(s) and mix the powder in molasses or peanut butter. The goat may eat the molasses willingly or you can scrape the peanut butter onto the roof of the mouth and let the goat clean it up.

2. **Oral liquids** - Administering an oral liquid to a goat (drenching) is often a very messy job. You will end up wearing some of the liquid, your goat will end up wearing some of the liquid, and hopefully, your goat will have swallowed some of the liquid. Several different instruments can be used to administer liquids to a goat - a dose syringe, a regular syringe (without a needle), and a *plastic* turkey baster. The turkey baster is probably the most readily available and is easily replaceable if damaged. The most important aspect of dosing a goat or sheep is to keep the head on a level plane (**Figure 43**). If the animal absolutely refuses to swallow the medication, it may run out onto the ground. But if you tip the head back to help hold the liquids in the mouth, you run the risk of the goat

accidentally inhaling some of the fluid. Inhalation of liquids will usually cause pneumonia and pneumonia is far more serious than the loss of some of your medication. If you infuse the liquid slowly and steadily with the turkey baster, your goat or sheep will usually swallow with only minimal complaint.

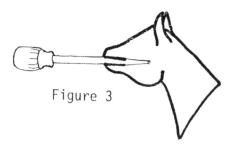

Figure 3

3. Injections - All small ruminants will

Figure 43. Drenching

require injections more often than oral medications. Most vaccinations are given by injection. Oral antibiotics may do major harm to the beneficial rumen bacteria whereas injectable antibiotics cause less problem.

Giving an injection starts with filling the syringe. Most vaccinations use small doses so a 3cc syringe with a 22-23 gauge, 1 inch needle is the appropriate size to use. (Smaller diameter needles hurt less when penetrating the skin. The smaller the diameter of the needle, the *larger* number when discussing gauges.) Many antibiotics are quite thick and require doses according to the weight of the animal. A 6-20 cc syringe with a 20 gauge, 1 inch needle will usually suffice. Needles that are 18-19 gauge are really too large for routine use in goats.

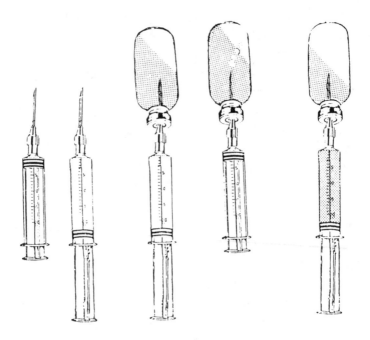

Figure 44. Filling the syringe

When filling a syringe, first make sure the needle is firmly attached to the syringe. Then pull the plunger back and fill the syringe with the same amount of air you intend to use of the injectable material. Push the needle into the bottle and inject the air into the bottle. Then pull the plunger back and refill the syringe with the vaccine or antibiotic. (**Figure 44**) If you don't replace the material removed from the bottle with air, a vacuum will eventually be produced and it will become very difficult to get the remaining material from the bottle into the syringe.

Injections can be made into many areas of the body. Most medications used by the "average" herd manager are given intramuscularly or subcutaneously.

Subcutaneous (SQ) injections are given into the loose tissue between the skin and the muscles (**Figure 45**). They are most often given over the shoulders or in the axillary region since these two areas tend to have a looser skin

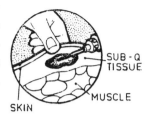

Figure 45. Sub-Q

attachment. See **Figure 47** for locations of subcutaneous injection sites. The skin is picked up with one hand and a "tent" is formed between the skin and the body. The syringe is held parallel to the body and the needle is pushed forward into the space in the tent. The hold on the tent is then relaxed and the injection is made. If the plunger is very difficult to push, you have probably left the hole in the end of the needle *in* the skin

(intradermal) by either not pushing it in quite far enough or by pushing it too far and almost coming out the other side (**Figure 46**).

Figure 46. Intradermal

Rearrange the needle slightly and then continue with the injection.

Intramuscular (IM) injections are given into the middle of a muscle (**Figure 48**). There are nerves, arteries, and veins in muscles so some care must be taken to avoid these. **Figure 47** shows the preferred sites for IM injections in goats and sheep with the small numbers listing the sites in order of preference.

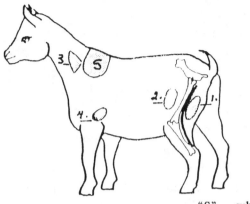

"S" = subcutaneous
#1-4 = intramuscular

Figure 47. Injection sites

Since Site #1 is most commonly used, I'll discuss it in a little more detail. The shot will be given in the muscle mass behind the femur bone. You should have an assistant (or a stanchion) hold the goat's head. Then stand beside the goat on the opposite side you intend to use for the injection. Lean over the animal and grasp the muscle mass in your hand. This mass should be moveable unless you have grabbed the

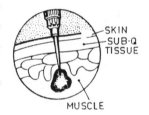

Figure 48. IM

femur along with it. With the syringe aimed in from the side (not the rear of the animal) insert the needle into the middle of the muscle mass in your other hand. Pull back on the plunger (aspirate) and be sure that no blood comes back into the syringe. If there is no blood, go ahead and give the injection. Otherwise, pull the syringe out and begin again. IM injections are often painful, so be sure to have a good grip on the goat, the syringe, and the muscle mass. Try not to let the goat's kicking move the needle too much in the muscle. If the syringe is too close to the femur, or if it is aimed in from the rear of the goat, there is a possibility that the needle or injected material may damage the sciatic nerve. This could cause temporary or permanent paralysis of the leg.

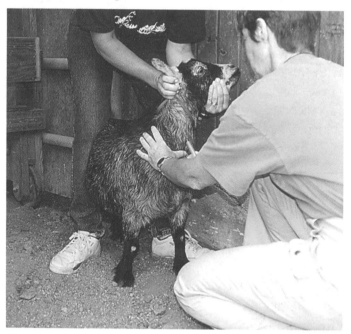

Figure 49. Intravenous injection

Intravenous (IV) injections are given directly into the vein and should usually be administered by your veterinarian. Most drugs that require IV use are very irritating to the muscle or subcutaneous tissue and it is very important that all the solution enter directly into the vein and not into the area around the vein. Faster drug actions are also obtained by IV administration of those drugs. Some drugs should never be given IV and could even cause fatalities when given by that route. Follow label directions carefully.

Occasionally you will notice a small lump appear at the area of a SQ injection given several days previously. It is possible that your needle was dirty and a small abscess has formed. But it is more likely that your SQ injection was given intradermally (into the thin layer of the skin rather than in the tissue layer below the skin) and the skin is reacting to the drug. Some of these reactions will eventually open and drain; others will go away very slowly over the course of several months. These kinds of small reactions to injections are a nuisance but will cause no health problems with your animal.

---

cc = cubic centimeter
ml = milliliter
mg = milligram

For all our usage, "cc" and "ml" can be used interchangeably to indicate the amount of material to measure in a syringe. "Mg" is a measure of weight and the number of milligrams per cc is different for most drugs. Do not confuse milligrams with milliliters.

DISEASES

GASTRO-INTESTINAL PROBLEMS

Teeth

Goats are browsers. They prefer to eat leaves and bushes. The goats' tooth structure is different than dogs, cats, or humans and compliments their eating habits.

Mammals in general have four classifications or kinds of teeth. Incisors are the smaller teeth in the front of the mouth. Next in line are the canines, which are also called fangs or eye teeth. Goats do not have any canine teeth. Premolars are the larger teeth immediately behind the canines and right behind them are the molars. Scientists use a "dental formula" to indicate the number of each classification of teeth a particular species has. The right and left sides of both the upper and lower jaws are mirror images of each other, but often the top and bottom jaws will differ in number of teeth. The dental formula for the goat is

$$\text{right and left sides} \quad 2 \left(\begin{array}{l} \text{upper jaw \&} \\ \text{lower jaw \&} \end{array} \; I\frac{0}{4}, C\frac{0}{0}, P\frac{3}{3}, M\frac{3}{3} \right) = 32$$

I = incisors, C = canines, P = premolars, M = molars

This dental formula tells us that goats have no canine teeth and that they have no incisors on the upper jaw. If you hadn't already noticed that, take a minute right

123

now and go look behind the lips of your calmest goat. You should find 8 lower incisors (4 on the left and 4 on the right) and no upper incisors. Goats have a dental pad instead. It is a very firm structure covered with tough gum tissue and should oppose nicely with the lower incisors (see **Figure 51**).

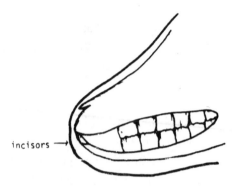

incisors →

Figure 51. Even bite

Goats do most of their chewing with their premolars and molars (also called cheek teeth since they are all hidden behind the cheeks). These teeth are hypsodont (high crowned or ever-growing) teeth and keep growing even in adult animals. As the top surface wears away, the tooth continues a slow eruption so the goat won't wear it's cheek teeth down to the gum from too much chewing. As the goats chew their food or chew their cud, they tend to wear the teeth unevenly and end up with slanted teeth from inside to out. The upper teeth may form sharp edges which irritate the inside of the cheek (see **Figure 52**). Some goats will treat themselves for this painful condition and will stuff some cud material between the teeth and the cheek. Eventually, quite a pocket may form in the cheek and the animal may appear to have an abscess or infected tooth but usually will not appear to be in any distress. The owner or veterinarian can more effectively treat the condition by"floating" the teeth. A small flat file is inserted in the mouth and the rough edges of the teeth are filed smooth. This involves filing the outside edge of the upper teeth and the inside edge of the lower teeth. Unfortunately, if the goat has already developed the bad habit of carrying a cud in her cheek, she may continue to do so even if there is no clinical reason.

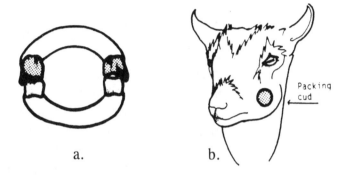

Figure 52. a. Uneven molar wear b. Packing cud

The continual growth of the cheek teeth can also cause problems if a goat loses a tooth. Occasionally a tooth will break or become abscessed and will need to be extracted. The animal will recover from the initial extraction quickly, but 6 months to 1 year later will begin to lose weight and not eat well. The tooth which opposed the now missing tooth will have continued its growth with no wear from the other side and will have become long enough to interfere with proper mastication (see **Figure 53**). If a tooth must be extracted or is lost for any reason, you must either extract the healthy opposing tooth or file the opposing tooth monthly to mimic normal wear.

125

Figure 53. Overgrown cheek tooth

Incisor teeth are brachydont (low crowned) teeth and do not continue to grow once fully erupted. The gums will slowly retract from the roots as the animal ages so the teeth will appear to be longer. Remember that the teeth are wearing off, too, so the lengthening may not be very visible. They will also tend to loosen somewhat and will lean farther forward in the older animal (see **Figure 54**). This forward inclination is called proclination.

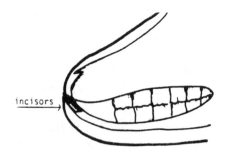

Figure 54. Old age

Goats' teeth are also indicators of age. This brings up

the fact that goats have baby teeth and permanent teeth just like we do. The dental formula for the baby, or deciduous, teeth is

$$2 \left(Di\frac{0}{4}, Dc\frac{0}{0}, Dp\frac{3}{3} \right) = 20$$

(D = deciduous)
There are no baby molars.

Figure 55 illustrates the approximate eruption dates for all the deciduous and permanent teeth. But for practicality, the cheek teeth are very difficult to examine and are not often used in aging goats. The incisors are quite accurate up to 4 years of age. The deciduous and permanent incisors look similar; the only real difference is size and the permanent incisors are quite a bit larger than the deciduous ones. Goats celebrate each of their first four birthdays with a new pair of incisors. So if you notice one or both of a pair of incisors loose or missing, don't panic. New ones are not far behind. When the new incisors are coming in, they often are turned partially sideways. But since I have only rarely seen crooked, fully erupted teeth in goats, I don't worry while they are erupting. Once all the incisors are permanent, aging becomes more difficult. Wearing of the incisors should be indicative of age past 4 years, but we tend to spoil our goats and their teeth. We hand them a flake of hay and a bowl of grain which requires little, if any, use of incisors for eating. We can't, therefore, expect much wearing of incisors, and examination of older goats shows this to be true. Long, loose or missing incisors are more indicative of old age than well-worn incisors. But let's keep spoiling the goats and their teeth. I know my goats' ages by their registration papers - who needs teeth?

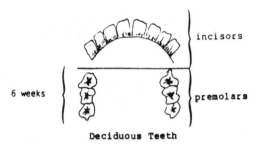

Deciduous Teeth

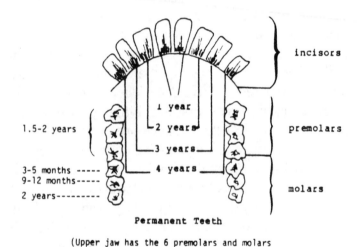

Permanent Teeth

(Upper jaw has the 6 premolars and molars
to correspond - for a total of 32 teeth)

Figure 55. Eruption dates

Eruption Dates

birth - 1 week	Di 1 erupted
2 weeks	Di 2 erupted
3 weeks	Di 3 erupted
4 weeks	Di 4 erupted
6 weeks	Dp 1,2,3 erupted
3-5 months	M1 erupted
9-12 months	M2 erupted
1-1.5 years	I 1 erupted
1.5-2 years	P 1,2,3 erupted
2 years	I 2 erupted / M 3 erupted
3 years	I 3 erupted
4 years	I 4 erupted (full mouth)
6 years	I worn and falling out (broken mouth)
8 years	I lost (gummer)

THE RUMINANT STOMACH

Pygmy goats are ruminants, as are sheep, cattle, deer, llamas, etc. Their digestive tract looks and functions quite differently from the simple stomached animals, such as man, swine, dogs, cats, etc. Knowing some of these differences can be very important for achieving the best herd management.

The ruminant stomach has 4 compartments, a rumen, reticulum, omasum, and abomasum. The simple-stomached animal has a single compartment stomach.

The newborn ruminant functions as a simple-stomached animal. At birth, the rumen is small and

129

non-functional. When the kid nurses, a band of muscle tissue called the esophageal groove closes to form a direct tubular connection from the esophagus to the abomasum. If you are bottle feeding kids, it is important that you hold the kid in a natural nursing position so that the esophageal groove closes properly. If milk is put into the rumen, either by tube feeding or improper bottle feeding, a considerable time may elapse before the milk makes it to the abomasum where it can be digested for use. Milk may also curdle in the rumen and cause some colic symptoms.

As the kid matures and nibbles on his environment, he slowly introduces the microorganisms (bacteria) which are necessary for proper functioning of the rumen. Kids raised by their mothers develop their rumens more quickly than bottle babies - presumably due to their mothers' influence on their eating habits. Very few bottle babies observe their two-legged mothers eating hay!

Each of the four compartments of the stomach has a particular purpose and function. The rumen is the largest chamber of the ruminant stomach (1-2 gallon capacity) and has no digestive enzymes. It is a large fermentation vat populated by microorganisms which change non-digestible cellulose into proteins which can be used by the body. Roughage is initially chewed and swallowed and goes to the rumen. Here it is worked on by the microorganisms, is regurgitated as a cud, and is rechewed. The process is repeated again and again until eventually the chewed food passes on through the reticulum to the omasum in a condition ready to be digested by normal body enzymes. Methane is continuously produced as a by-product of the bacterial action, so eructation (an odiferous belch) is a sign of a healthy rumen. Anything which harms the rumen microorganisms can effectively halt the digestive process.

The reticulum lies in front of and below the rumen near the liver. Its lining is honeycombed and it serves as a catch chamber for heavy articles in the feed. In cattle, magnets are often placed in the reticulum to catch and hold the various nails, pieces of wire, etc. that the animal might swallow. Since goats are more fastidious in their eating habits, magnets are not necessary. The reticulum from cattle is sold in the market as tripe.

The omasum is divided by long folds of tissue which help decrease the size of food particles coming from the rumen and which also help remove excess fluid.

The abomasum is the true stomach and is the only compartment which produces digestive enzymes. It acts on food prepared by the rumen just like the simple stomach acts on food entering from the mouth. It is the area of primary digestion of all grain and milk. Grain and milk do not require the efforts of the rumen microorganisms.

While being a ruminant does allow our pygmies to thrive on a diet of hay, it also leaves them open to many different diseases which affect the 4 chambers of the ruminant stomach. Knowing a bit about the stomachs will help understand those diseases.

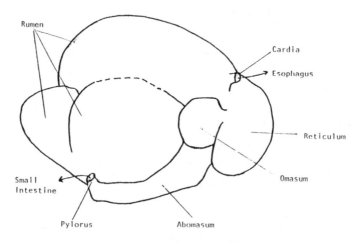

Figure 56. Ruminant "stomach" compartments

INDIGESTION

A healthy rumen is mandatory for proper digestion of roughage in the goat. The digestive enzymes in the abomasum and small intestine cannot work properly on roughage unless it has been prepared by the rumen microorganisms. So, for the goat to stay healthy, its rumen bacteria must stay healthy. These bacteria can be killed by improper feeding (too much grain, moldy hay or grain, dog food, pig food, etc.), oral antibiotics, and pathogenic bacterial toxins (such as those produced by *Clostridium perfringens, type C & D*). When the bacteria die, the rumen becomes a vat of decaying food and bacteria which quickly becomes toxic to the goat. The toxins released by dead and dying bacteria are absorbed by the goat. The rumen fluid becomes very acidic which causes the goat's blood to become too acid. The rumen quits its natural contracting and becomes stagnant, so more bacteria die and the process continues. This "indigestion" can range from mild to very severe to fatal.

Treatment of indigestion is simple in theory. The rumen must be detoxified, encouraged to contract and empty, and must be restocked with normal bacteria. Milk of magnesia will detoxify and reduce the acidity of the rumen. It will also help encourage rumen contractions. The dosage for an adult pygmy goat is 2 ounces (1/4 cup) given four times a day. Keep in mind that milk of magnesia will NOT cause diarrhea in an adult ruminant. However, as all the toxic material is emptied out of the rumen, there will be a foul-smelling diarrhea for 12-24 hours. Do not treat this diarrhea; you want to clear all the bad materials as quickly as possible. To replace normal bacteria, the old time farmers recommend a "cud transplant." But far easier than stealing a cud from a healthy goat is buying cultured yogurt or acidophilus tablets at the store. One-half cup of yogurt daily will act very well as a cud transplant. If the milk of magnesia and yogurt routine does not work, your vet will want to give intravenous fluids to combat the acidosis and shock and injectable drugs to stimulate the rumen to start its contractions. There is a definite point of no return when the rumen quits, so it is not wise to delay treatment. In fact, if you find your goat with her head in the grain barrel, you would be wise to start the milk of magnesia treatment immediately. She will already be starting to recover by the time she first shows signs of being sick. If you wait until she acts sick to start treatment, you may not be successful.

Most goats do not like milk of magnesia or yogurt. It is best to put on old clothes and to use a plastic turkey baster to dose the goat. By the time you are through, both you and she will be wearing the milk of magnesia and she hopefully will have swallowed some, too.

If a specific bacteria, *Clostridium perfringens, type C or D*, is present in the rumen, it takes advantage of the acid environment produced in indigestion and multiplies rapidly, producing its own toxins. This disease is enterotoxemia or over-eating disease. It is only one possible cause of indigestion, but is usually quickly fatal and does not respond well to any treatment. Enterotoxemia can be prevented by vaccination. All animals on pastures and at risk of getting into the grain shed should be vaccinated. These animals will be most susceptible to the disease. Young nursing kids are also at risk, especially if their dam is producing lots of milk. They may be normal at night and be found dead in the morning. Vaccination is indicated if your herd has had this problem in the past. Goat kept on "dry lot" conditions with absolutely no chance of getting excess grain may not need this vaccine.

BLOAT

The normal rumen churns 1 to 4 times every minute, and its bacteria produce methane gas continuously. Most of this gas is released as the animal eructates (see **Figure 57**a.) Certain diets, especially fresh, green alfalfa, will cause the gas to form tiny bubbles which are entrapped in the rumen fluid. This is called frothy bloat. The tiny bubbles cannot be released in a natural belch and the condition progresses rapidly until the rumen is grossly distended and the animal is extremely uncomfortable. Position of the animal may also be a cause of bloat. If a sick or anesthetized animal lies on its side, the cardia of the rumen (the opening between the rumen and esophagus) will be low and the natural gas pocket in the rumen will be above it (see **Figure 57b**.) The gas is again trapped and the rumen becomes painfully distended.

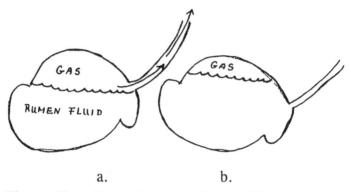

a.   b.

Figure 58. a. Normal gas position   b. Bloat

Treatment rationale here is obvious - the gas must be allowed to escape. The old time farmers would "stab" the animal high in the left flank into the gas pocket in the rumen and release the excess gas. Rarely, if ever, are such heroics required with goats nowadays. Position the goat on a steep incline with the front legs higher than the rear (see **Figure 58**.) This artificially elevates the cardia and will often be all that is necessary the relieve a positional bloat. Mineral oil or milk of magnesia (2-3 ounces) will help with a frothy bloat by breaking the tiny bubbles to form one large gas pocket which can be relieved normally. Once the drugs have been administered, massaging the abdomen and walking the goat will help with proper mixing and breakdown of the bubbles. Relief from frothy bloat should be evident within one hour of administration of the drugs. If these treatments do not work, your vet may need to pass a stomach tube to release the pressure in the rumen while the drugs have a chance to work.

Figure 58. Positioned to relieve bloat

Do not confuse a full rumen with bloat. The rumen lies on the left side of the animal and a true bloat will cause a tense, firm swelling in the left flank and the animal will be in obvious distress. However, some animals will eat a big meal and will look bloated. but they are comfortable and can easily belch and/or bring up a cud.

CHOKE

Occasionally a goat will eat something which obstructs its esophagus, an expanded pellet or a piece of carrot or apple. The goat will drool profusely and if the obstruction is complete, it will become bloated since the rumen gas has no way to be released. However, the treatment in this case must be directed at removing the obstruction. Most often, skillful passage of a stomach tube will relieve the choke and the bloat simultaneously. In extreme cases, an esophagostomy may be performed if the cause of the choke can be palpated in the neck.

Prevention is directed at stopping the rapid ingestion of the pellets or grain. Spreading them out in a large pan may be all that is necessary. Occasionally it is necessary to put some large smooth stones in the pan to make the goat eat more slowly.

RUMEN IMPACTION

If a goat eats too much dry roughage or not enough water, the rumen may become impacted. Once the rumen is overstretched, it is difficult for it to contract. As the liquid is absorbed from the roughage, the problem intensifies. Palpation of the left flank will reveal a firm, doughy rumen. The goat will not be chewing a cud but occasional belching may be seen. The goat will act hungry, but will eat only very small amounts. Only a small quantity of feces will be passed.

Treatment is directed at rehydration of the goat and the rumen contents and at increasing contractions of the

rumen.   In mild cases, giving extra water by mouth
may be sufficient treatment. Milk of magnesia may also
help lubricate the rumen contents and stimulate
contractions. Fresh green feeds will also help increase
rumen activity. Avoid grains and dry foods. Serious
cases will require intravenous fluids, oral drenching,
and drugs to increase contractility. As a last resort, a
rumenotomy may be performed.

ABOMASAL IMPACTION

Abomasal impaction is most common in young kids.
They may have grain introduced but will go for many
days eating little or none of what is offered. Then, one
day they decide that grain is great and they overeat.
The abomasum becomes overfilled and the kid goes off
feed. Treatment is the same as for rumen impaction.
Rehydrate the animal and the food in the abomasum and
increase gastric contractions.

DISPLACED ABOMASUM

The abomasum belongs in the most ventral portion of
the abdomen lying on the abdominal floor. In very rare
instances, the abomasum will rotate and become
positioned along the left abdominal wall between the
wall and the rumen.   It can also rotate the other
direction and end up along the right abdominal wall. A
right sided displacement is an actual torsion and causes
acute obstruction, severe pain, and high mortality. A
left-sided displaced abomasum (LDA) causes only a
partial obstruction with a decrease in volume within the
abomasus and interferes with normal contractions. The
goat will go off feed and have a decreased volume of

soft stools or diarrhea. A mild ketosis will occur. Treatment is usually surgical replacement and tacking the abomasus to the abdominal wall.

HARDWARE DISEASE (Traumatic Reticuloperitonitis)

Hardware Disease is uncommon in goats and good management will keep it uncommon. Cattle tend to eat almost anything, including nails, staples, and pieces of wire. These sharp objects stay in the reticulum but they may puncture through the reticulum wall and migrate into the chest or abdomen. Hardware disease is the complex of problems caused by the migrating sharp objects.

Goats have far more discriminating appetites than cows so are less likely to eat sharp metal objects. (They are much more likely to eat plastic, which can also lead to serious or fatal intestinal impaction.) Well kept pens with no loose nails or small pieces of baling wire will help prevent problems. Any goat with vague abdominal distress, sudden anorexia, reluctance to move, pain during defecation and urination, and a low grade fever should have Hardware Disease included in the differential diagnosis. They will also have decreased rumination and a leucocytosis with a left shift.

Treatment may be conservative with parenteral antibiotics and fluids and confinement of the animal for at least two weeks. Roughage in the feed should be decreased. More aggressive therapy is rumenotomy and removal of the offending object from the reticulum.

INTESTINAL OBSTRUCTION

Despite their discriminating appetite, goats will readily eat plastic and cellophane. Small pieces and quantities will cause no harm in the rumen, but may cause an obstruction at the sphincters between stomach compartments. If it does get into the intestinal tract, it may obstruct anywhere along the length of the small intestine. The goat will go off feed, and stools will decrease in quantity and may become loose. Treatment is removal of the blockage. Surgery will usually be required to achieve this.

ANTIBIOTICS

There is no arguing that antibiotics are "wonder drugs." Without them, our goats (and humans and other animals, too) would be dying from bacterial diseases. Antibiotics either kill or halt the growth of bacteria and allow the body to recover from any damage that was done.

There are two kinds of bacteria that are important to animals. Pathogenic bacteria are those which cause disease and non-pathogenic bacteria are those which either help the body or at least peacefully coexist with it. But antibiotics are indiscriminate in their actions and will kill off necessary intestinal and ruminal bacteria as well as the pathogenic bacteria. This is a very important side-effect in ruminants since ruminal bacteria are mandatory for proper digestion. The magnitude of effect on ruminal bacteria will vary with the particular antibiotic used and with the duration and dose of the drug.

Watch for loss of appetite or depressed attitude in any ruminant that is on antibiotic treatment. Treat immediately with medication to replace rumen bacteria. My old standby treatment is yogurt. Yogurt contains natural Lactobacillus acidophilus to help re-populate the rumen. Acidophilus capsules may also be used. And there are commercial products such as Probiocin® that contain natural bacteria and vitamins in a palatable formulation made specifically for ruminants. One of these products should be used during or following any long term antibiotic therapy and with any short term therapy if the animal shows clinical signs of rumen upset.

DIARRHEA

Diarrhea is defined as abnormal frequency and liquidity of fecal discharges. Normal goat stool is passed as small, roundish pellets. A very early stage of diarrhea is manifested as a clumping together of the pellets with the individual pellets still easily distinguishable. One large solid stool is the next step and it progresses to less formed and more watery. Diarrhea is not an immediate health concern unless it is watery and there is a risk of dehydration and electrolyte imbalance. The earlier stages are warnings to figure out what the problem is before it becomes serious.

There are many possible causes of diarrhea. In young kids, bacterial infection, coccidiosis, and other internal parasites are the most common causes. Enterotoxemia will cause diarrhea in weanlings and adult animals. Overeating and fresh greens (pasture, grass clippings) are common causes in more mature animals.

Don't immediately treat all goats for diarrhea. Depending on the cause, sometimes it is best to allow the disease to run its course for a day or two unless there is a risk of dehydration. Sometimes it is best to "flush out the evil demons". Kids that are not yet ruminating (<6 weeks) can be treated symptomatically with kaopectate or Imodium. Specific treatment for the primary cause is also indicated. Adults of all ruminating animals, should receive milk of magnesia and yogurt. Milk of magnesia will NOT cause diarrhea in adult ruminants. It does detoxify, and the yogurt helps restore some normal bacteria. If the animal is being fed grain, decrease the amount until the diarrhea is resolved.

JOHNES DISEASE(Paratuberculosis)

Johnes Disease is caused by *Mycobacterium paratuberculosis*. Goats are infected as kids by ingesting stool from infected adults, but the disease has a 1-5 year incubation. The bacteria damages the intestinal lining and causes malabsorption and a wasting disease. The animal becomes anemic, weak, and develops a poor haircoat. Diarrhea may or may not be a symptom. Some goats will go the entire course of the disease with normal stools.

Diagnosis can be made by blood test (AGID tests) or fecal culture. Unfortunately, the culture takes ten plus weeks before results are available. A fecal smear can be stained with an acid fast stain. A positive stain leaves Johnes as a possible cause of disease, but a negative stain may only mean that no organisms were found in that particular smear. Some goats may be inapparent carriers and may not be shedding virus at the time of the tests. These goats will test negative by culture.

Treatment is not effective and the disease is contagious. There is no vaccine currently available in the US. Prevention is best achieved by isolating suspected carriers, especially from young kids. It is also wise not to allow kids to nurse from any animals that test positive on the blood test. Animals that are positively infected should be removed from the herd.

COCCIDIOSIS

Coccidiosis is a form of internal parasitism caused by several species of protozoa (single-celled organisms). There are many different species of coccidia with at least three that can affect goats. Coccidia tend to be very species specific–they affect only one species of animal. A chicken with coccidiosis is not infectious to goats.

Coccidiosis is spread by fecal contamination of feed and water. Since it is endemic in goat herds in the United States, all our goats will be exposed sooner or later. In most herds, the older animals have all been exposed and are carriers. They show no clinical signs of disease, but are passing coccidia in their stools. This provides a continual source of infection in their pens. Young kids then ingest contaminated feed and become infected too. Protecting feed and water from contamination with fecal material will help decrease the incidence of Coccidiosis, but we all know that young kids are very inquisitive creatures and will always search out the few flakes of hay or pieces of weeds on the ground before eating fresh, unpolluted hay from the feeder. So unless you can prove otherwise, assume that your herd is infected and treat it accordingly.

The clinical signs of Coccidiosis can vary from no signs at all to sudden death. Some animals will become infected, recover, and become immune carriers with never a visible sign of illness. Other animals die suddenly from hemorrhage into the intestinal tract while still passing normal stools. In between these two extremes, goats may show a depressed appetite and mild diarrhea or severe bloody diarrhea. Kids may show a dry, rough hair coat and a bloated belly. Some recovered animals will have permanent damage to the intestinal tract and may not grow and develop properly.

A diagnosis of Coccidiosis is easily made by examining a fecal sample. The parasite will be found in high numbers in most samples. Occasionally illness will occur before the parasite begins to be excreted in the stool. In these cases, treatment must be initiated on the basis of clinical signs only; or a diagnosis may be made at necropsy. Limited numbers of parasites (up to 3/low power field) should not be considered diagnostic of clinical Coccidiosis. While Coccidiosis occurs primarily in young kids, it may occasionally take advantage of an adult with other intestinal problems. It should be considered as a possible diagnosis in all cases of diarrhea.

Coccidiosis can be treated with a course of sulfa drugs or amprolium. Sulfadimethoxine (Albon®) given for 5-7 days will clear up most cases. If a recheck fecal exam one week after treatment still shows high numbers of the parasite, either a repeat course of Albon or a course of amprolium is indicated. Amprolium (Corid®) can be used at two different dosage levels. The higher dose, given for 5 days, is used to treat infected animals. The lower dose, given for 21 days, is used as a preventive treatment for animals likely to become infected. Amprolium must be dosed properly since overdosage can cause a vitamin B_1 deficiency.

Prevention is the byword of successful goat breeders. This is especially true with Coccidiosis since a less than successful response to therapy may leave an unthrifty animal with a scarred intestine. Once you are aware that you have Coccidiosis in your herd, a little carefully timed treatment can prevent any signs of disease. Treat all 4-5 week old kids with a course of Albon. Bottle babies can be more prone to developing clinical Coccidiosis. It may be advisable to treat these kids at 2-3 weeks of age and again at 6-8 weeks. It is not necessary to run a fecal exam first. Waiting for the appearance of parasites in a microscopic examination may well be waiting until clinical disease is already underway. One course of treatment at this time is usually all that is necessary. Of course, clean pens and feeders built to avoid fecal contamination (which means feeders built to keep kids out) are also a very important step toward successful prevention of Coccidiosis.

Decoquinate (Deccox®) has been approved for the prevention of coccidiosis in young goats. The dosage is 0.5mg per kilogram per day. This dose should be mixed in the feed and fed for at least 28 days during periods of exposure. It should not be fed to breeding animals or animals producing milk for food.

PROLAPSED RECTUM

A prolonged course of diarrhea or constipation may cause a rectal prolapse. Chronic coughing may also be a cause. The abdominal straining forces the rectum to "turn inside out" and part of it then protrudes through the anus. Minor cases may prolapse only when the goat is defecating or when lying down; and then it returns to its proper anatomical position when the animal moves or stands up. More severe cases do not reduce themselves automatically. Three to four inches stay protruded and will be damaged by straw, other goats, etc.

Treatment involves reduction of the prolapse and prevention of immediate recurrence and elimination of the inciting cause. Reduce the prolapse by cupping it in your hand and applying steady pressure. The goat will push against you, but firm, constant pressure will overcome the resistance. Once reduced, insert a finger into the anus to make sure the tissue inside is "straightened out." If the prolapse keeps recurring, a purse-string suture needs to be placed in the anus. It should be tight enough to retain the prolapse while still allowing normal defecation. An occasional goat will need to have some stool manually (digitally) removed. The suture should remain in place about 7 days.

MUSCULO-SKELETAL PROBLEMS

LACERATIONS

Even with the best barn management, accidents will happen and goats will get injured. Punctures should be cleaned and kept open for 5-7 days and local antibiotics should be applied. If tetanus vaccination is not up to date, a booster should be administered. Small cuts should be treated similarly to punctures, but may require sutures. Larger wounds usually deserve veterinary advice. Deep cuts should be flushed and pressure should be applied if they are bleeding. A sanitary napkin works well as a clean pad in a pressure bandage. Do NOT use blue lotion or other non-washable products until a decision is made regarding the necessity of suturing.

ARTHRITIS

Arthritis is the inflammation of a joint. There are many causes of arthritic conditions in goats. Most will appear similar clinically, but treatment regimens and prognosis will differ depending on the cause.

Infectious Arthritis can be caused by Corynebacterium, Staphylococcus, Mycoplasma-Chlamydia, and other bacterial infections of joints. These organisms may spread to the joint through the blood or they may enter through a puncture into the joint or surrounding tissue. One or more joints will become acutely inflamed and

147

the animal may have a fever. A culture and sensitivity of joint fluid is definitely indicated, but some organisms are very difficult to grow. It is always best to take fluid for culture PRIOR to the administration of any antibiotics. Intra-articular injection of appropriate antibiotics may speed recovery. This injection can be made at the time the fluid is taken for culture. Some organisms are not particularly sensitive to antibiotics and the disease process may slow down but not stop. These animals will have chronic lameness. Mycoplasma and Chlamydia are probably transmitted through colostrum and/or milk. Clinically unaffected and clinically recovered animals may be carriers and can transmit these organisms.

Joint Ill is a specific form of Infectious Arthritis caused by a bacterial infection which enters through a fresh umbilical cord and spreads to one or more joints. Three-four week old kids are most commonly affected, but the disease may appear at any time from 1-12 weeks of age. The kid will become suddenly lame with enlarged, tender joints and will have a high body temperature. The carpal and elbow joints are often the joints affected. Antibiotics are the treatment of choice. A culture and sensitivity of joint fluid would give a definitive diagnosis and would indicate the most effective drug to use. A gentamicin and ampicillin combination is a good choice while awaiting culture results. Warm compresses on the affected joints will help ease the animal's discomfort. A small amount of exercise or physical therapy is indicated to maintain as good a range of motion as possible. The kid will be very painful during the early course of this disease, but most will respond well. In some cases, low doses of corticosteroids may be indicated to alleviate some joint inflammation. Some will be left with some permanent joint damage and chronic lameness. Others will recover with no ill effects. A few animals will not respond to treatment and will require euthanasia.

Trauma may also cause arthritis. Usually only one joint will be affected. The original trauma may cause a lameness due to soft tissue damage which appears to recover completely. Degenerative arthritic changes may appear at a later date and will be insidious in onset. Treatment is symptomatic. Aspirin can be given for inflammation and pain. It works best given twice daily for about 5 days to stop the inflammation. This 5 day course may need to be repeated periodically. Prevention of the arthritic condition may best be accomplished by proper treatment of the initial injury. Splinting or support wraps may be needed to help protect the joint cartilage while the soft tissue heals.

Improper nutrition of young growing animals, bucks especially, may also cause arthritic changes in joints. Calcium, phosphorus, and Vitamin D imbalances are the most likely causes of Nutritional Arthritis. If the disease process is caught early, and the diet is corrected, any arthritic changes may be stopped and possibly even reversed.

TETANUS

Tetanus is caused by *Clostridium tetani*, an anaerobic bacteria. This bacteria is found in the soil, so is readily available to contaminate wounds. *Clostridium tetani* grows in the absence of oxygen. Deep puncture

149

wounds from dog bites, nail punctures, etc. that close over are excellent places for the tetanus organism to grow. Elastration and disbudding wounds are also excellent growth media. Animals in the process of teething may contract tetanus as the teeth loosen. Superficial wounds, large open wounds, and clean surgical incisions are unlikely to allow tetanus organism growth.

Clostridium tetani produces a neurotoxin that travels along the nerve sheath to the spinal cord. By the time clinical signs are seen, the neurotoxin is well disseminated. The incubation period from the original injury is 1-3 weeks. The animal will have a normal temperature and will be fully mentally aware. The first signs of tetanus are difficulty chewing, swallowing, and picking up food. The animal can still walk but will be stiff and will over react to external stimuli. The ears become very erect and may cross over the to top of the head. The third eyelid may protrude and the animal will tend to assume a sawhorse stance. As the disease progresses, the animal becomes more rigid and lays on its side with its neck hyperextended (see **Figure 59.**) Death eventually occurs from cardiac or respiratory failure.

Treatment involves good nursing care and high doses of tetanus antitoxin and penicillin. Remember that even the totally rigid animal is still mentally aware. He is distressed that he has no control over his body. Keep him in a dark, quiet area. Use tranquilizers to help achieve some relaxation for the goat. Clean and debride the original wound if possible. Administer tetanus antitoxin intravenously one time. Give penicillin daily. Intravenous fluids and tube feeding may be indicated.

Figure 59. Tetanus in a kid

The prognosis for goats with tetanus is very poor. Always give tetanus prevention with any possible anaerobic wound. Vaccinate all kids and keep those vaccinations current.

FOOT ROT

Foot rot is caused by *Bacteroides nodosus*, an anaerobic bacteria. This organism grows well in a wet environment and is highly contagious. The goat becomes lame on one or more feet. The skin between the hooves becomes necrotic. The hoof walls are undermined and loosen from the hoof proper. The foot becomes very foul smelling.

Treatment is aimed at drying the foot and eliminating the necrotic tissue. Get the goat in a dry environment. Trim the feet as radically as possible to remove all dead tissue. Soak the feet in Coppertox® after trimming. You may want to make a foot bath of zinc sulfate that the goat must walk through daily.

Foot rot will usually respond well to treatment if it is begun early in the course of the disease. As the disease progresses, it becomes more refractory to treatment. The infection may spread to the bones in the foot and cause permanent damage. Affected animals do spread the organisms.

A vaccine is available for use in sheep. It does cause a large local reaction at the injection site. Cleanliness and dry barns may be a more appropriate means of control for most pygmy goat herds.

LEG INJURIES

Leg injuries can occur no matter how careful you are. Check corrals and stalls periodically for possible problems. Goats can catch their legs in holes or notches in the fence, forks in trees, hanging brackets, etc. They are very imaginative at figuring out ways to injure themselves.

A sudden onset of lameness with or without a swelling on the leg is often indicative of a traumatic injury. Injuries may vary in severity from mild sprains or hematomas to severe fractures.

Treatment will depend on the nature of the injury. The biggest worry is often breakdown of the normal leg. Goats don't do well on three legs. A fractured leg may heal well, but if the goat was 3-legged for many weeks, the normal leg may show signs of joint damage from abnormal usage.

CAE

Caprine Arthritis-Encephalitis (CAE) is an ongoing problem in goat herds. CAE is a specific and important disease and contributes to the early demise and/or crippling of many goats. It has also become the "scapegoat" category for many undiagnosed problems.

CAE is caused by a retrovirus. This virus is carried in milk and colostrum and is readily absorbed by a young animal which drinks the milk. The virus does not cross the placental barrier and clean kids are born from infected does. There is evidence that there may also be transmission from goat to goat after prolonged (1 year or more) direct contact.

Exposure to CAE virus causes the formation of specific antibodies in the blood and these antibodies can be detected by a blood test. Approximately 80% of the nation's goats were shown to be positive on this test in the first year it was run. Most of that 80% will never show any clinical signs of disease, but may be able to pass the virus on to their offspring. About 10-20% will develop signs of arthritis or encephalitis.

CAE develops in two primary forms in pygmy goat herds. Young animals tend to exhibit the neurologic signs associated with encephalitis. Weakness and incoordination begin in the hind legs and progress to include the forelegs. Kids do not run a fever and remain bright and alert. Most do not survive the disease. Some kids may show signs of arthritis or pneumonia along with the encephalitis. Older animals, usually over 1 year, will develop swollen knees (see **Figure 60**), stifles, and/or hocks with a slowly progressive lameness. They lose body and coat condition. The adult form of the disease may progress slowly over the course of many years or may progress more rapidly within 1-2 years. Pneumonia, wasting, and udder edema may also occur with this virus.

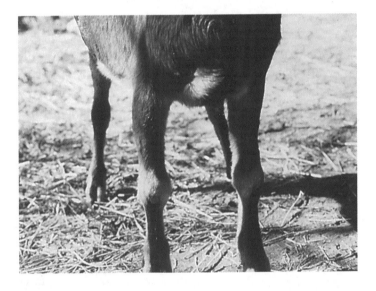

Figure 60. Swollen knees from CAE

Due to the nature of the virus, there is no quick cure for CAE. The prospects for development of a vaccine are not good. Treatment is routinely unsuccessful and consists primarily of supportive care and pain control. Prevention through management, however, is quite feasible - if you're truly committed to limiting this disease in your herd. There is still a lot we do not know about this disease, however, and some cases will seem to appear out of nowhere even in the best and most careful management programs.

Since the primary mode of transmission of CAE is through the ingestion of contaminated milk, it is only reasonable that feeding only uncontaminated milk should halt or hinder its transmission. However, in most pygmy herds, this will be a major management commitment.

Steps for control of CAE virus:

1. Blood test herd - the first step in prevention is blood testing your herd and identifying all positive does. Milk from these does should not be fed to any kids unless it is pasteurized.

2. Feed only milk from negative does or pasteurized milk. You must be present at all deliveries of positive does to ascertain that the kid does not nurse. One good nursing of infected colostrum is all that is needed to infect a kid. Kids must then be bottle fed clean milk. Several years of this management, even allowing for one or two surprise deliveries with contented, nursing kids, should leave you with a high percentage of negative yearlings and kids.

3. Cull all positive animals. At this point, you would ideally begin to cull older positive stock.

Horizontal transmission (from mature goat to mature goat) is less common, but does occur. If you don't wish to cull heavily, you might consider the separation of positive older animals from negative and younger ones. Do NOT test kids under 6 months of age. They may exhibit a false positive test due to colostral antibodies.

4. Periodic rechecks of the herd for CAE virus are wise. Negative animals may convert to positive as late as 5-6 years of age.

These management procedures are a lot more work for the breeder/herd manager than the natural process of does feeding and rearing their own kids. If your herd has little or no arthritic or encephalitic disease problems, then it's probably too much extra work with no noticeable beneficial result.

Remember that CAE is extremely widespread. Approximately 10% of the goats in the United States will develop clinical signs of CAE. Absolute proof of the diagnosis of CAE is nearly impossible for the veterinarian in the field. Many animals with the diagnosis of CAE have an entirely different but clinically similar disease. Both you and your veterinarian must consider all possibilities before assigning the diagnosis of CAE to an animal, and even then, you must consider it a probable rather than a definitive diagnosis.

Management and herd health programs can only be determined on an individual herd basis. What's good for one herd may not be applicable for the herd just down the street. Evaluate the importance of CAE in your herd. If it is a major problem, then consider the management procedures discussed here. But if it is not a problem, then why tamper with success?

SELENIUM

Selenium has long been a controversial subject in the veterinary community, medical community, and among breeders and owners of animals. One faction touts the beneficial effects of selenium while the other side stresses the toxic effects of overdosage. Where is the happy medium? Is there a right answer?

Selenium is a trace mineral. Trace minerals are required by the body in small, but specific, amounts. Too little will cause deficiency symptoms and can be fatal; too much will cause toxicity and can also be fatal. The safe zone may be small, but is does exist. And it behooves us all to find that safe zone for our small ruminants.

Selenium is available to animals as an inorganic salt and in organic forms in alfalfa, grasses, etc. Alfalfa has a low capacity to extract selenium from the soil so that animals fed mainly on alfalfa are more prone to be deficient in selenium. In certain areas of the country there is an excess of selenium in the soil and care must be taken to avoid toxicity. Some forms of selenium are "unavailable" to the animal. Even though it eats a large measurable quantity of selenium, it is not able to be absorbed by the body. Therefore, measuring the selenium content of your goat or sheep's feed will not answer all your questions regarding supplementation unless you can determine the bioavailability of the

157

mineral.

While selenium deficiency is a problem in the majority of the United States, some of the Great Plains states have the opposite problem. They are very rich in selenium. Certain plants are "accumulators" of selenium. Locoweed and milk vetch are two of the plants which may contain enough selenium to be toxic. But both these plants have an unpleasant odor and are unpalatable to most livestock. Animals must be forced to eat such plants (by not furnishing enough palatable feed). The management of most goats I have seen leaves this type of poisoning very unlikely.

Numerous clinical disorders of goats are now associated with selenium deficiency. These include White Muscle Disease, retained placentas, infertility, slowed growth, unthriftiness, and other related problems. Usually young animals are most obviously affected, while older animals may have equal problems which are not as clinically apparent. Selenium deficiency classically affects the muscular system, although more recent evidence indicates liver, gastrointestinal, and reproductive involvement as well.

White Muscle Disease (Nutritional Muscular Dystrophy), the most well known disease caused by deficiency of selenium, is characterized by a muscular dystrophy involving leg and/or heart muscle. Animals affected from birth to 3 months of age may show difficulty in rising and unsteadiness when standing or walking. Those animals with affected heart muscle may look clinically like pneumonia but will not respond to treatment for pneumonia. Older animals may have weak pasterns as their only sign of deficiency. Some animals just don't grow properly and this is usually evident by 2 to 4 months of age. This lack of proper growth is often accompanied by a disorder in liver function due to selenium deficiency. Pregnant females are more readily affected by a selenium deficiency since they must supply selenium for both themselves and their fetus(es). Some dams will fail to implant an embryo; some will resorb the embryo at an early date if there is a lack of selenium. Others keep the selenium for themselves and deliver weak or dead babies. Still others give the selenium to the babies and become deficient themselves. These animals may have difficulties with delivery due to lack of uterine tone to expel the kids properly or possibly just lacking the tone to involute the uterus properly after delivery and to expel the placenta.

At the other end of the spectrum, selenium poisoning can be either acute or chronic. The acute form shows an animal with sudden death or labored breathing, blindness, abnormal movement and posture, diarrhea, prostration, and death. The chronic form, when the animal is poisoned a little at a time over a length of time, is called "Alkali Disease."

Alkali Disease is characterized by dullness, emaciation, stiffness and lameness. In cattle, there is loss of hair at the base of the tail and swollen coronary bands and deformities of the hoof. Sheep do not show foot and wool lesions like a cow, but they do exhibit a reduced rate of reproduction. There is no published data on selenium poisoning in goats. Watch for any and all the signs listed for sheep and for cattle. Young animals are more susceptible to selenium poisoning than are older ones.

Selenium deficiency is best prevented before clinical problems occur. There is trace mineralized salt available with selenium added. There are injectable selenium compounds which, in most areas, should be administered to all breeding animals twice a year (if the selenium salt is not used). Many newborn kids should also be treated. Animals already showing clinical signs may be treated with the same injections and most will recover quickly and fully. A few animals which are treated late in the course of the disease will be left with some permanent muscle disorders, but they will not continue to deteriorate.

Diagnosis of clinically apparent selenium deficiency is made on clinical signs and response to treatment. In suspicious cases, lab analysis of blood will usually show elevated muscle, and occasionally liver, enzymes. Animals treated with selenium injections (Bo-Se®) may show a marked improvement in 2-5 days. When this occurs, I treat again at one week and at one month. Then the animal is put on a twice yearly schedule. If no response is seen, the diagnosis should be reconsidered.

Selenium poisoning can be difficult to diagnose. There are some laboratories that are able to run a serum or hair analysis for selenium. Absolute normal values are still somewhat questionable. There is no specific treatment for selenium toxicity. Removal of the animals from the toxic feeds is definitely indicated in both prevention and treatment.

Vitamin E is very closely related to selenium in its actions within the body. Both Vitamin E and selenium should be used in treating selenium deficiencies. There is some thought that the current preparations of selenium and Vitamin E are not balanced properly. If you are quite suspicious of a selenium deficiency and do not get a proper clinical response with your treatment, you might try some additional Vitamin E.

Selenium is added to most goat feeds with the maximum amount being regulated by federal law. Discuss the advisability of supplementing your goats with selenium with your veterinarian. At the same time, do not overdo your supplementation. The labels on the selenium bottles are the dosages approved by the Food and Drug Administration. A lot of research has been done by the drug companies to obtain this approval. Remember that many of the signs of selenium deficiency are not readily visible (decreased reproduction) so don't bury your head in the sand and think that your herd does not have a problem. The greater portion of the US does have a deficiency.

REPRODUCTIVE DISORDERS

VAGINAL PROLAPSE

On a routine examination of your does, or ewes, you notice that one doe who is due to deliver in four weeks has a "blob" of pink or red tissue protruding from her vulva. Do you panic? Of course not. You are an informed herd manager. You think through the possible reasons for this abnormally placed tissue and then you calmly decide how to properly care for the doe.

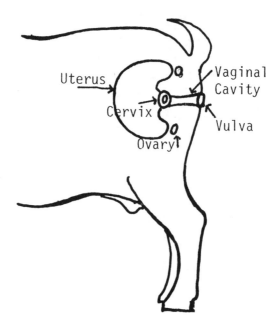

Figure 61. Normal anatomy

There are two immediate diagnoses that come to your mind.

1. vaginal prolapse
2. uterine prolapse

There are always other highly unlikely possibilities, but you are content to consider only the most probable ones first. You catch the doe in order to examine the "blob" more carefully - but it disappears when she stands up. (A clue!)

A vaginal prolapse usually starts with a swelling of the floor of the vagina. In most does, the swelling will protrude from the vulva only when the doe is lying down and will be the size of a golf ball or small orange (see **Figure 62**a.) When the doe stands up and takes a few steps gravity tends to pull the tissues back inside the vulva. These prolapses require NO treatment.

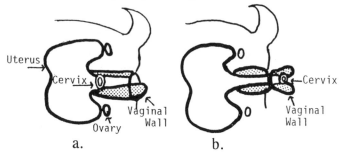

Figure 62. a. Moderate prolapse b. Severe prolapse

Slightly more serious is the prolapse that is small grapefruit size when the doe is lying down. These do not easily reduce themselves when she stands up, and there is concern that the prolapsed tissue will be scraped or torn. The diligent herd manager will assist the doe by manually reducing the prolapse as often as feasible during the day. First, gently cleanse the prolapsed tissue with warm water. (Do not scrub the already traumatized tissue.) Cup your open hand around the tissue and exert a slow steady pressure to replace it. Once the tissue has gone back behind the lips of the vulva, insert one or two fingers into the vagina to make sure that all the vaginal tissue is back in its natural position. Holding the swollen tissue in place for 30-60 seconds will help keep it in place.

Most serious is the prolapse that involves the entire circumference of the vagina and that refuses to stay reduced (see **Figure 62b**.) The exposed vaginal tissue becomes quite swollen and thickened. These prolapses must be manually reduced and measures must be taken to force them to stay reduced. For years, this required suturing the lips of the vulva shut. The tissue would stay slightly swollen but was limited as to how much swelling could occur and external trauma was prevented. However, the herd manager must be present to cut the sutures prior to delivery or severe damage could occur to the doe and possibly the kid.

In sheep, there has been a second alternative. A device called a bearing retainer (available from Jorgensen Laboratories) is inserted into the vulva (see **Figure 63**.) The sides of the retainer are tied to the wool to keep the device from falling out. The long tongue of the retainer holds the cervix in place and prevents any swollen vaginal tissue from leaving its rightful place inside the vagina. When the ewe is ready to lamb, the cervix dilates and the lamb can be delivered right alongside the bearing retainer.

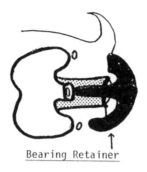

Bearing Retainer

Figure 63. Bearing retainer inserted

The bearing retainer sounds like the perfect answer, but the goat doesn't have any wool to hold the device. The ingenious herd manager must rig up an elaborate harness to take the place of a few pieces of wool. A horse halter can be used as a harness and should fit closely but not too snugly. One long length of 1 inch sewing elastic is then passed through the hole in the end of the bearing retainer (already positioned in the doe). One end of the elastic is then crossed over the top of the goat and attached to the harness at the shoulder on the other side. The other end passes behind and to the inside of the rearleg and attaches to the harness at the middle of the chest floor. The two top elastics should be tied together where they cross on the top. The lower pieces each require another elastic to attach to the lower piece just in front of the hindleg and stretch to the top elastic just in front of the bearing retainer. The elastic should be barely stretched. This will keep the retainer in place and will still allow the doe to lie down and arise without difficulty. When delivery time comes, the initial bag of amniotic fluid will stretch the elastics and push the retainer out of the vagina. Since the

165

prolapsing tissue surrounds the tongue of the retainer and is not pushing from the end of the tongue, it will not push the device out of position.

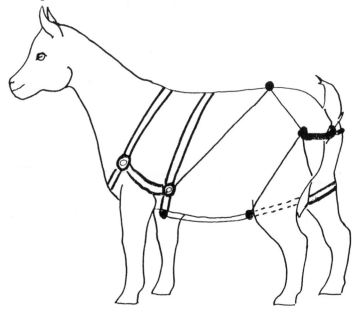

Figure 64. Harness for bearing retainer in goat

Vaginal prolapses usually occur a week or two prior to delivery, but may occur as much as 4 weeks prior. The time between the initial diagnosis and the time of impending delivery will make a difference in how much concern the herdsman must have. There is one nice feature of the less serious vaginal prolapses. When the doe goes into labor (8-12 hours prior to actual delivery), the prolapse will reduce itself and will stay reduced. As the cervix dilates and the uterine tone is increased, the pull of the internal tissues keeps the vaginal floor in place. Only rarely will a doe have trouble with delivery because of a prolapsing vagina.

Uterine prolapses (see **Figure 65**), on the other hand, are extremely serious and life endangering. They usually occur immediately after delivery. The entire uterus turns itself inside out and hangs out of the vulva behind the goat. This type of prolapse must be cleaned and reduced as soon as possible and will require suturing to hold the tissues in place. Contact your veterinarian immediately if this occurs.

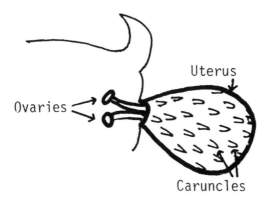

Figure 65. Uterine prolapse

Vaginal prolapses are apt to recur in the same doe year after year. The swelling of the vaginal floor may be due to the increase of estrogen being produced in preparation for delivery. It is possible that exogenous sources (from outside the body) may be responsible for some of the increase of estrogen. Legumes, such as alfalfa, may have high quantities of estrogen. Other possible causes of vaginal prolapse are increased abdominal pressure in late pregnancy and coughing.

Now that you have mentally gone over all the possibilities of what could be occurring with your doe, you are content to leave her with the rest of the herd and just keep a close watch one her. You know that the prolapse may become worse in the next month before she delivers, but until it does, you will leave it alone.

DYSTOCIA

Dystocia is defined as abnormal labor. You must understand the normal delivery process before you can tell what is abnormal. It is important to be present when the delivery begins so you are aware as it progresses.

Allow only 1/2 - 1 hour of hard labor before getting or providing assistance, but don't get over-enthusiastic. If the kid is part way delivered and then gets stuck, wait a few minutes. If the shoulders are stuck, be sure that the sack is broken off the face because the kid does need to breathe; his cord is clamped in the birth canal. Once he is breathing, he can stay in this position for a long time, if necessary to drive to a veterinarian. If the kid is a posterior presentation and the shoulders are stuck, it is more of an emergency to get the kid delivered. His cord is clamped in the birth canal, too, and he may try to breathe and drown in amniotic fluid.

There are many causes of dystocia and each must be treated appropriately. The most common cause is malpresentation. The logical treatment for this is to correct the presentation and pull the kid. Common sense, a good sense of feel in you fingers, and the ability to create a mental picture of how the kid is positioned will help accomplish this. The object in correcting a malpresentation is to rearrange the kid to a normal anterior or posterior presentation, or, at least close enough to be deliverable.

The possible list of malpresentations is endless. Just when I think I've seen them all, a kid manages to arrange himself in a totally new way. Twins, triplets, etc. are even more complicated since several can try to exit at once. Clean your hand and arm well. Use an obstetrical lubricant or leave a bit of soapy water on your arm and reach into the doe. Feel the kid to decide what you are feeling (and that it is all from the same kid) and form a mental image. Now decide which body part, leg or head, to find and pull into the birth canal to allow a delivery. Try to avoid pushing the kid back into the doe. Once the uterus has contracted enough to push the kid part way out, it is easy to tear if too much pressure is applied back against it.

Always turn the kid very carefully. If you are trying to pull a leg around, cup the hoof in your hand as you turn it. The hooves do have a soft covering while in the uterus, but can still tear a uterus if they are handled incorrectly.

If you try to assist a doe by turning the kid and are not successful within 10-15 minutes, then take your hand out of her and call for the veterinarian. Do not try for 30-60 minutes. You will wear out the doe and cause internal tissues to swell. The job will be harder for the vet, and the kid and doe will be at greater risk. Let the doe rest while awaiting more experienced help.

After you have pulled one kid, plan on pulling the rest. If you rearranged the kid quickly and pulled it with minimal trauma to the doe, then give her 10-15 minutes to contract her uterus and work on the next kid herself before helping again. If the first kid was hard work to get out, then only wait a couple of minutes before pulling additional kids. The doe will be very tired and the remaining kids may start to have some oxygen deprivation as the placentas start detaching. Once the last kid is out, put a uterine bolus (an antibiotic formulated specifically to dissolve in uterine fluids) into the uterus to help prevent infection. Be sure the goat's tetanus vaccine is up to date. Depending on the difficulty of the delivery and the condition of the kids, injectable antibiotics are also indicated. Do NOT pull the afterbirth.

Dead kids can be difficult to deliver. They are not well lubricated and do not flex well during the delivery process. They may be presented correctly and still require pulling.

Malformed kids may or may not be able to be pulled.

Kids that are overly large for the doe will not be able to be delivered vaginally. A Caesarian section will usually be required.

Some does will have uterine inertia and the uterus will not contract. Calcium and oxytocin may be administered with care. Goats are quite sensitive to calcium; it must be given very slowly intravenously while monitoring the heart or it can be given subcutaneously. Oxytocin should not be given unless you are sure that the cervix is fully dilated and the kid is properly presented.

Rarely, you may miss the signs of stage 1 and early stage 2 labor. The doe may attempt to deliver unsuccessfully and then close down her cervix without delivery. A Caesarian section will be required to deliver those kids; their chance of survival is slim.

Does with chronic selenium deficiency may have very weak uterine musculature. Labor initiates properly and the cervix dilates, but the uterus cannot push adequately to expel the kid. If you reach in to examine the doe, you may feel like you are reaching into a cavern. Give some oxytocin, pull the kids, and give Bo-Se. Repeat the oxytocin in 12-24 hours and place a second uterine bolus at that time too.

RETAINED PLACENTA

If the afterbirth is not delivered within 1-5 hours, don't panic. Do NOT pull the afterbirth or try to disconnect individual cotyledons. You may cause serious or fatal hemorrhage. If it does not pass within 24 hours, call your vet. Most often antibiotics will be administered and the placenta will liquefy and be passed over 4-5 days. Placentas are more often retained with premature deliveries and induced deliveries.

TORN UTERUS

On very rare occasions, a uterus may tear during a natural delivery. More often, a uterus will be torn during an assisted delivery. Pulling a kid with an incompletely dilated cervix may cause a tear. Fetal hooves have a soft covering to help protect the uterus, but the maneuvering during assisted repositioning for delivery puts more than normal stress on the uterine wall. Always cup the hoof in your hand when pulling on legs and changing the orientation of the hoof. Care must also be taken if a kid is repelled during an assisted delivery. The uterus is a very strong muscular organ that is pushing hard to expel the kid. Any effort made to push the kid back into the uterus must be done slowly and carefully by an experienced person. If too much pressure has to be exerted to repel the kid, then the procedure should be aborted and some other method of achieving delivery should be pursued.

A doe with a torn uterus will be in pain. But she may not show clinical signs for several hours or more after delivery. An untreated tear will lead to severe peritonitis and death. A very small tear may heal itself with minimal peritonitis but there may be problems with future pregnancies. Either the doe won't settle or she may abort early.

Tears can be diagnosed by direct palpation of the uterus

through the vagina. If the tear is in a horn of the uterus or in the cranial portion of the body, surgical repair via laparotomy can be accomplished. If the tear is near the cervix, access for surgical repair is difficult. Following surgery, pain medication such as Banamine® is indicated. Antibiotics should also be administered. If the kid will be drinking the milk, it is important to use an antibiotic that will hot harm the kid.

PROLAPSED UTERUS

Sometimes a doe will not be content to just deliver her kids. After the last kid is born, she will continue to push and will turn her uterus inside out and end up with the whole uterus hanging out the vulva. The prolapsed tissue will often be covered with dirt and manure. Venous drainage from the uterus will be compromised so it is important to reduce the prolapse as soon as possible. The placenta(s) may or may not still be attached. If the placenta had already detached, the caruncles on the surface of the uterus will be easily visible. If the placenta is still attached, only detach it if it comes easily. Forcible separation of cotyledonary attachments may lead to severe hemorrhage. Clean the prolapsed tissue off and place a clean sheet under it. Raise the prolapsed tissue up level with or slightly above the vulva to help reduce the edema in the tissue. Slowly and steadily push the prolapsed tissue back into the vagina, being careful not to perforate the tissue with your fingers. The doe will tend to have contractions during this procedure. It is not unusual to almost completely reduce the prolapse and then to have the doe push the whole uterus back out in a moment when you relax your pressure. Some does contract so much that the prolapse cannot be reduced. These does must be tranquilized or anesthetized. An epidural anesthesia will work very well. Once the prolapse is reduced, it is imperative that the inside of the uterus be palpated to ascertain that all "corners" of the uterus are back in their proper anatomic position. If one is still inverted into the uterine cavity, it is more likely to try to prolapse again. The lips of the vulva should be sutured to prevent a re-prolapse. Sutures can be removed in 7-10 days. Oxytocin should be administered after this procedure is completed to help contract the uterus back to normal size and position.

Figure 67. Prolapsed uterus

Uterine prolapse in not common. It occurs most frequently following a difficult delivery. It is not likely to recur with subsequent deliveries.

METRITIS

Metritis is an infection of the uterus. This occurs most frequently following difficult or assisted deliveries but may also occur in any doe. Various causes include unclean assisted deliveries, ascending infection from a natural delivery, ascending infection during estrus. Clinical signs include fever, vaginal discharge, anorexia, and weight loss. The doe may go down and refuse to stand. Chronic cases may be due to retained dead or mummified fetuses. These may not be diagnosed for 6-12 months. The doe may lose weight and will not have normal estrus cycles. There is usually no vaginal discharge since the cervix remains closed.

Treatment involves emptying the uterus and controlling the infection. If the infection has occurred immediately after delivery, an injection of oxytocin will expel uterine contents. At most other times, prostaglandins are more likely to help empty the uterus. Antibiotics are best chosen on the basis of culture and sensitivity of uterine discharge.

KETOSIS (Pregnancy Toxemia)

Signs of ketosis are caused by a decreased blood glucose and increased ketones. It occurs in the last 4-6 weeks of gestation, most frequently in does with multiple fetuses and in overly fat or very thin condition. These animals do not handle carbohydrate metabolism correctly. Either there are inadequate nutrients for the doe and her fetuses, or excess fat along with multiple fetuses take up so much abdominal space that the doe

175

no longer eats adequate amounts.

Does affected with ketosis will go off grain and then off hay. They become uncoordinated, hold their head elevated, and are weak. As the disease progresses, they develop a mucoid nasal discharge, rapid breathing, and will not stand. They can hold themselves in sternal recumbency. Blindness may be exhibited early or late in the disease. Muscle tremors may develop, followed by coma and eventually death.

Treatment in early cases is propylene glycol. Give 2 ounces twice daily for 2 days. More severe cases should receive glucose intravenously. If the response is poor, labor should be induced or a Caesarian section performed. Once the kids are delivered, the drain on the doe is decreased dramatically and her metabolism will return to normal. The outlook is not good if the doe was unable to stand prior to surgery.

This disease is better prevented than treated. Keep your does in good weight and provide proper nutrition in late gestation. If ketosis has been a problem in the past, provide a molasses supplement during the last month of gestation. Begin treatment immediately if you are even suspicious of impending ketosis.

MILK FEVER

Milk fever is caused by a low blood calcium. It is rare in goats, but it can occur just prior to or in the first few weeks after delivery. Clinical signs develop abruptly. The doe becomes stiff and uncoordinated with the rear legs being more severely affected. She will develop muscle tremors, weakness, and apprehension. Eventually she will lie on her chest, unable to rise, with her head extended forward and hindlegs extended backwards. Paralysis, coma, and death follow. The body temperature remains normal or low throughout this process.

Treatment is calcium given very slowly intravenously. If the doe is diagnosed early in the course of the disease, give the calcium subcutaneously. Goats are very sensitive to calcium.

Milk fever can be prevented with good management. Feed a diet such as oat hay that is low in calcium for the month prior to delivery. This helps mobilize the body's calcium stores so that it can respond to the increased demand with parturition and lactation. Supplementing with calcium during the month before delivery will defeat this purpose. Switch back to alfalfa just a few days prior to freshening.

MASTITIS

Mastitis is an inflammation of the mammary gland usually caused by infection. It can occur under the best of circumstances, but more often is due to improper milking procedures. Mastitis can occur quite suddenly in an acute infection or may slowly decrease production and damage the gland in subclinical or chronic infections. A bacterial infection may be introduced into the udder during or immediately following milking. Cleanliness during milking and teat dipping following milking are very important methods used to prevent mastitis.

If your doe has developed mastitis in spite of your efforts to prevent it, you must first diagnose, and then treat it. With acute mastitis, the udder will be hard and hot and may be very difficult to milk. Only small amounts of milk may be released and this will probably be an abnormal secretion. The milk may be watery and may have small clots of tissue in it.

Ideally the milk should be cultured to find out specifically what bacteria is causing the problem and what antibiotic will be effective against it; but often this is not economically or physically possible. Under any circumstances, the mastitic udder should be milked as frequently as possible during the day to keep the bacteria milked out and to leave as little milk as possible in the udder for further growth of the bacteria. After your last milking at night, an antibiotic infusion (labeled for mastitis in cows) should be placed in the udder to help kill the bacteria. Before infusing the mastitis ointment, the udder should be emptied and the end of the teat disinfected with alcohol. Then with a gently twisting motion, insert the end of the mastitis ointment tube into the teat canal and infuse one-half of the medicine into each side of the udder. After withdrawing the tube, hold the end of the teat closed and massage the medicine up and out of the teat into the udder. If your doe has very small teat canals and the infusion tube will not fit through the opening, just invert the end of the teat slightly with pressure from the tube and then expel the medicine - most will go into the teat canal.

Usually 3 days of frequent milking and 3 nights of antibiotic infusion will clear up mastitis and you can then proceed with normal milking. You should not drink mastitic milk, and you should not drink milk for several milkings after using a mastitis ointment. The exact number of milkings to be withheld from human consumption will be listed on the mastitis ointment tube. When you prepare to dry this doe up, she should be treated with a "dry cow" mastitis ointment. Dry her up as usual, but two weeks later, empty the udder and infuse one-half a tube of dry cow preparation into each side of her udder. This will help prevent her from carrying an infection through to her next freshening.

BRUCELLOSIS

Brucellosis is not a common problem in goats. However, it is of public health significance since it can infect people either by direct contact with infected goats or by ingestion of contaminated milk products. The disease is called Malta Fever in man.

Brucellosis is caused by *Brucella melitensis* and causes abortions which are seen primarily in the last two months of gestation. This disease is transmitted by ingestion of infected aborted materials. The fetus, fetal membranes, and fetal fluids all contain the Brucella organism.

Testing for Brucellosis is required for most interstate transportation of goats. It is also strongly suggested to test any doe whose milk is to be consumed without pasteurization.

CYSTIC OVARIES

Goats are usually bred on a regular basis and their reproductive tract is kept functioning in a natural manner. Many pygmies are kept as pets and are only bred once or twice or possibly not at all. These animals that aren't regularly reproducing are more likely to develop cystic ovaries, although they may occur in reproductive animals, too.

Occasionally, follicles will develop on the ovary which do not rupture, or ovulate, properly. They keep producing estrogens to bring the doe into season, but still don't ovulate. Often, these does come into heat every 5-8 days continuously. (This is different from the 5-day heat discussed in Hormonal Control of the Estrus Cycle, page 59) They may exhibit some masculine characteristics, become more aggressive, mount other does, etc. They also may produce milk, even if never bred. These does usually require an injection of Cystorelin® or lutenizing hormone to force ovulation of the cystic follicles and start normal cycles once again. If she is being kept as a pet, she may require surgical removal of her ovaries to return her to an acceptable behavioral state. If cystic ovaries are undiagnosed or untreated for too long, they may not respond to treatment and the doe may be a loss to your breeding herd. Untreated or unsuccessfully treated cystic ovaries may lead to bone marrow suppression.

URINARY PROBLEMS

URETHRAL OBSTRUCTION

Male goats, especially wethers, are at risk of urethral obstruction from small bladder stones (see **Figure 68**) trying to pass through the urethra. The goat's penis is long and has an "S" shaped curve (sigmoid flexure) built into it. The urethra, through which the urine passes, is small in all males and smaller in wethers. Stones form in the bladder. When the goat urinates, some stones are expelled from the bladder with the urine. Very small stones will pass unimpeded, but they may cause some irritation to the lining of the urethra. The irritated urethra swells and stones become trapped, especially in the sharp turns of the sigmoid flexure or in the urethral process at the tip of the penis.

The obstructed goat will strain to urinate, exhibit discomfort when laying down, and occasionally cry in pain. Early in the course of the disease some urine may leak around the obstructing stone and drips of urine may pass from the end of the penis. Dampness and occasional crystals may be found around the opening of the prepuce. Treatment with anti-inflammatory drugs and smooth muscle relaxants may enable to goat to pass these stones. As the disease progresses, the urethral tissue swells around the stone and no urine can pass. The bladder fills to over normal capacity. If the obstruction is not relieved, either the bladder or the urethra will rupture to relieve the pressure. This is painful and often fatal. It is easy to mistake the straining goat as constipated, and all too many blocked

181

goats are improperly treated with laxatives. Pay attention to your goat. Observe if he has urinated.

Figure 68. Urethral stones

Treatment of the obstructed goat requires removal of the obstruction. If the stone is in the urethral process, it is just cut off. If stones are caught at the sigmoid flexure, they must be removed surgically. If the obstruction cannot be relieved, a perineal urethrostomy can be performed. This surgery will create a new urethral opening under the goat's tail and bypass most of the penis. This surgery will eliminate the breeding capabilities of the goat. Unfortunately, strictures are a frequent complication of this surgery. If there are stones remaining in the bladder, a cystotomy to remove these stones should be performed at the time of the urethrostomy.

Prevention of recurrence can be very difficult. Analysis of the stones will help guide dietary changes. Maintenance of proper calcium-phosphorus ration, encouragement of water consumption, and urinary acidification can be pursued.

DERMATOLOGY

CASEOUS LYMPHADENITIS (Boils)

At one point, caseous lymphadenitis (CL) was the most common caprine disease in the United States. Luckily, it has been less common in pygmy goats than in dairy goats.

It is caused by *Corynebacterium pseudotuberculosis*. This organism has a thick outer wall and survives well in the environment. It can enter the body through lightly abraded skin. It is picked up by phagocytic cells and survives their destructive enzymes. The phagocytic cells congregate in the lymph nodes. The corynebacterium organisms outlive the phagocytes and multiply in the nodes. The node becomes abscessed. Seventy-five percent of the abscesses occur in the head and neck area of the goat.

Goats with CL have abscesses in lymph nodes (see **Figure 69**.) The pus is quite thick and may be white, yellow, or tan. Most goats will exhibit one or more external abscesses before any internal ones develop. An occasional animal will only develop internal abscesses, usually in the mediastinal and mesenteric nodes. These animals eventually show a wasting condition with no other visible signs.

Figure 69. Possible locations of abscesses caused by Caseous Lymphadenitis

External abscesses are not uncomfortable to the goat and rarely cause clinical signs. They are unsightly and may rupture. The pus contains live organisms and WILL spread the disease. It is best to surgically remove the smaller abscesses before they rupture. If surgery is not an option, isolate the goat BEFORE the abscess opens. Lance the abscess or allow it to rupture. Flush the open abscess with iodine and peroxide twice daily for 2 weeks minimum or until several days past the observation of any pus in the flushed material. The abscess must heal from the inside out. Do not allow the outside wound or incision to heal too quickly or the whole process will repeat. Large abscesses can be packed with gauze impregnated with iodine. Remove one-third of the packing each day. Do not return the goat to the herd until the wound is healed.

Prevention of CL is far easier than treatment. If you don't have abscesses in your herd, don't let them in. Check any possible purchases for palpable abscesses or scars from healed abscesses. Both these types of animals are probable carriers. There is no way to check for internal abscesses.

If you already have an abscess problem, try to slowly eliminate it. Isolate all animals with open abscesses; isolate them BEFORE they open. Prevent wounds and abrasions as much as possible. Check feeders and gates for sharp edges. Treat all wounds with disinfectants. Use only clean instruments for tattooing, disbudding, etc. Cull chronically ill animals; or at least consider isolating them from the rest of the herd.

Do not blame all abscesses on corynebacterium. Abscesses in non-lymph node areas are usually due to foreign bodies, infects wounds, or insect bites. They usually contain a very liquid pus. Abscesses in the brisket area are often mixed infections and are very difficult to treat.

Hint: Do not lance abscesses without tapping them with a needle first. Some lumps on the body are caused by hernias and should not be lanced.

SORE MOUTH

Sore mouth, also called contagious ecthyma, is a real nuisance to have in your herd, but it isn't as serious as many other diseases. It affects both sheep and goats and, occasionally, man. The disease is caused by a virus which is contained on the scab which covers the lesions on the mouth. The virus is highly contagious and is spread very similarly to small pox.

Figure 70 Sore mouth in a San Clemente Island goat

Sore mouth has a short incubation period of 2 to 3 days. The most common lesions occur on the mouth at the very edge of the lip. In more severe cases (see **Figure 70**) the lesions may spread up the skin of the muzzle to the nostrils. Some lesions may occur on the dental pad and the roof of the mouth. Occasionally, a nursing lamb or kid will spread the disease to the teats and udder of its dam. Extremely severe cases may involve the genitals and feet. The lesions you will see are small scabs approximately 1/8 to 1/4 inch in diameter. They

187

may occur right next to each other and appear to cover the entire lip edge. The scabs are tender to the touch and may crack when bent or moved (by the animal or the examiner). Within about 3 weeks, the scabs will dry up completely and fall off leaving newly healed skin underneath. The scabs will store the live virus for a long time (possibly up to 2 years). Those that fall off the animal as the disease progresses will then lie in wait to infect the next animal in the pen.

Treatment is essentially useless with this disease. Nolvasan ointment can be applied to severe lesions to help soften them and reduce the pain. Remember that the animals mouth is tender and difficult to open widely. Feed softer foods that are easily ingested. Coarse, stemmy hay will continually rub the lesions and cause pain. Leafy, fine stemmed hay will cause less local abrasion and will be far more comfortable for the animal to eat. Alfalfa and molasses, grains, and pasture will greatly help animals with more severe cases of sore mouth to maintain their condition while the disease runs its course. Animals that are forced to eat abrasive foods will crack the scabs and are more prone to secondary bacterial infection. These are the animals that are more likely to die from the disease. Careful management should prevent almost all deaths. However, a very occasional case will infect the inside of the rumen, digestive tract, and liver. These animals are not likely to recover even with the best care.

Next most important to careful feeding of infected animals is the prevention of spread of the disease. Ideally, if you find the problem in your herd, you keep all the animals that have come in contact with the infected animal in one pen so as not to contaminate other pens. You don't move those animals to a clean pen until their mouths are completely healed. Then they are immune to reinfection and are not contagious. If very many animals are involved, there should be a second pen for the healed animals to move to - just in case you move one out before it was ready. You can still avoid contaminating your entire property.

If the whole property has been contaminated, and if you show up with the problem again next year in new and newborn animals - you may have to vaccinate. This is one disease in which you may vaccinate even during the beginning of an outbreak. The vaccine is the live virus and you are actually giving each animal a limited case of the disease. (Remember your small pox vaccination? Sore mouth vaccination works on exactly the same principal.) The vaccine virus will also contaminate your property, so you don't really want to vaccinate unless you are trying to prevent a major outbreak in a large herd or flock.

Sore mouth is a nuisance. Your animals are essentially quarantined while the disease runs its course. You should not take animals off the property for sale or show. You should not bring outside animals in for breeding. Once the disease is over, you are safe to take animals off the premises but must still consider the contagion of scabs for any outside animal coming onto your property.

Orf is the name for this disease when it infects man. The disease is not highly infectious for man but will infect scratches and cuts that come in direct contact with the virus. If you have cuts or scratches on your hands, be sure to wear gloves when treating or examining any animals with sore mouth. The lesions are quite painful in man and take longer to heal than in the sheep or goat. Your doctor may have difficulty diagnosing your problem unless you suggest the possibility of orf or sore mouth.

ABSCESSES

Abscesses are the accumulation of pus. Treatment for skin abscesses involves opening them and flushing them well and assuring continuing good drainage.

Goats will occasionally develop a large abscess over the sternum between their front legs. They tend to chronically irritate the area and to keep it dirty each time they lay down. A protective "harness" can be made from an old canvas book bag. One of the handles is used as the strap to go over the goat's head. The rest of the bag is cut and re-sewn as shown in the diagram. A "maxi-pad" with an adhesive strip can be stuck on the harness to cover the wound and can easily be changed daily. A goat can wear this contraption for many months until the abscess has fully healed.

1. Slip neck strap over head.
2. Place bib over chest and between front legs.
3. Position belly band underneath and behind front legs.
4. Bring back strap up and over withers, inserting strap through slit in neck strap.
5. Bring back strap down and button onto belly band.

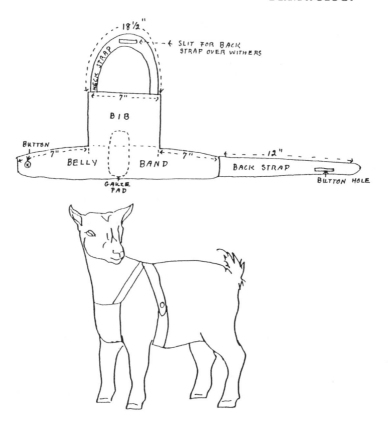

Figure 71. Harness for chest wound

RESPIRATORY PROBLEMS

SHIPPING FEVER

Shipping fever is the name given the respiratory disease which follows stressful situations, such as shipping, showing, etc. There are several viruses and bacteria which have been incriminated in this disease syndrome. There are vaccines available for the protection of cattle against this disease, but I have not found them particularly effective against the disease in goats. Usually, the goats arrive at the show looking super; they continue to look great during the show; and they often look good for the first day or so at home. Then someone gets a "snotty nose," and suddenly, all your goats who attended the show are coughing and have snotty noses. Within 2 to 3 more days, all the goats who had stayed at home are also affected. Often, the disease will get more severe as it works its way through your herd. The first animals may only require nose cleaning, and the last few animals may require antibiotic injections and lots of nursing care.

Is this preventable? The answer is "probably not." The viruses are passed through the air on water droplets that are coughed up by a carrier goat. Often the carrier may exhibit no signs of illness, but still may eventually get the disease right along with everyone else. The virus was there just waiting for its chance to attack while the goat was stressed.

Local show chairpersons can keep sick goats out of the barn and show ring, but they cannot keep out the viruses. So when you come home from a show and have to 'sweat out' the possibility of respiratory disease in your herd, don't blame the show chairperson, or another particular goat. The virus probably came with many of the animals, and maybe it was even there before the show - waiting for the next unsuspecting goat to come along. A long acting antibiotic given prior to traveling to a show and repeated just before traveling home may help prevent this disease. Flocillin or Liquamycin-LA 200 can both be used.

PNEUMONIA

Pneumonia can occur as an acute or a chronic disease. It can be caused by bacterial, fungal, parasitic or viral infections, allergies, or neoplasia.

The affected goat will usually have a cough. It may run a fever and be anorectic and depressed. Some goats will have difficulty breathing and most will worsen with exercise. Chronic cases will eventually exhibit signs of wasting. Upper respiratory signs such as nasal discharge are NOT often seen with pneumonia.

Diagnosis of pneumonia will require auscultation of the lungs and/or x-rays of the chest. To identify the specific cause, a tracheal wash with culture and cytology may be needed.

Treatment will depend on each specific case. Often the goat will need treatment either before culture results are available or before the owner is willing to pursue a specific diagnosis. History and signs may point to a likely cause and treatment may be administered accordingly. If there is inadequate response to treatment, then the lab results will be necessary.

Antibiotics are most often the first line of treatment. A drug should be used for 2-3 days before assuming that is isn't working. Any drug that does work should be administered for at least 4-5 days after clinical recovery. Nursing care is very important. The goat should be kept in a non-drafty and well ventilated area. In very cold weather, a sweatshirt or goat-coat will help.

BRONCHITIS

Not every goat that coughs has pneumonia. Some animals will react to a dry dusty environment or to dry, dusty hay with a mild bronchitis. These animals will behave normally. They will eat well and have a normal temperature but will have several episodes of coughing during each day. Most of these animals need NO treatment.

TUBERCULOSIS

Tuberculosis (TB) can occur in goats and the same organism can infect humans. Most states require a negative TB test prior to allowing movement of animals into the state. The organism can be shed in milk, so all animals producing milk for human consumption should also have a negative test or the milk should be pasteurized.

TB can produce mild to severe clinical signs. Chronic wasting and coughing may be the only signs seen. Other animals may have an intestinal form of TB with diarrhea as the primary sign. Diagnosis is made by intradermal skin test or by post-mortem exam.

NEUROLOGICAL PROBLEMS

POLIOENCEPHALOMALACIA

Polio is caused by a deficiency of thiamine (vitamin B_1) and is most commonly seen in 3 month to 2 year old goats. Thiamine is produced by healthy rumen flora. Some plants, such as bracken fern and horsetail, produce thiamine antagonists and will cause signs of deficiency even with adequate rumen production. Amprolium, a drug used to treat coccidiosis, may also cause this disease. Anything which upsets the rumen environment can change the level of thiamine production.

Goats will show an acute blindness, moderate to severe muscle tremors, and will have a normal body temperature. Often the muscle tremors are severe enough to cause convulsions. With no treatment the goat will die in 1-2 days.

This is a difficult disease to diagnose in the live animal. A good response to intravenous thiamine treatment is the quickest indication of a correct diagnosis. A severely affected goat will usually relax while the drug is being injected. Any goat with obscure neurological signs should receive thiamine as part of the treatment.

Polio usually occurs in isolated instances with no cause being discerned and with no affect on the rest of the herd.

196

MISCELLANEOUS

PINK-EYE (Infections Keratoconjunctivitis)

Pink-eye is a highly contagious disease thought to be caused by Mycoplasma and/or chlamydia organisms. There are numerous other possible causes of keratitis and conjunctivitis, however, most of these are not contagious.

The first signs of pink-eye are excess tearing from one or both eyes. The conjunctive becomes swollen and inflamed. The cornea of the eye will become cloudy or opaque and may have tiny blood vessels on its periphery. The eye is painful. If both eyes are severely involved, the goat may be blind.

Treatment will depend on the purpose of the affected goat. If the goat is intended for food production, either meat or milk, it should be treated with tetracycline eye ointment. If there is NO possibility that the animal will be used for food at any time in the future, then I prefer to use chloramphenicol with a corticosteroid as the eye ointment. It is illegal in California to use chloramphenicol in any food producing animal. The steroid is a controversial treatment in any eye with the possibility of corneal ulceration, but I have had excellent results with Ophthocort® or Chlorasone® in goats with severe pink-eye.

PAIN

Goats do not tolerate pain well. With many diseases or injuries, pain control is an essential part of treatment. Banamine works well for visceral pain. It should only be used for a couple of days. If longer treatment is necessary, then the diagnosis and treatment regimen should be re-evaluated. Aspirin helps with arthritic and other musculoskeletal pain. It may be given for up to 5 days to stop an inflammatory process and then will probably not be needed additionally.

POISONING

Plant poisoning can be fatal to goats and it is essential that owners identify toxic plants and prevent their goats from any access to them. Most well fed animals are less prone to problems with poisoning. This is partly because they are not ravenously hungry and, therefore, are more discriminate in what they eat. They also have a higher resistance to toxicity. But even the very best husbandry cannot prevent an occasional "accident" and subsequent ingestion of poisonous matter.

The early signs of poisoning depend on the toxin that was ingested. If you suspect anything, get to a veterinarian FAST. Early treatment is essential, and being able to identify the cause of the poisoning will help your veterinarian provide an antidote. Unfortunately, poisoning is often fatal.

The list of poisonous plants is long and varies in different parts of the country. A general rule is that all house plants, flowering bulbs, and ornamental foundation plantings (yew, azaleas, rhododendron, and laurel) are all poisonous. Wild cherry, bracken fern, foxglove, and rhubarb leaves are a few others on the list. It is wise to check with your local veterinarian and/or farm bureau for a complete listing of toxic plants in your area. Death by poisoning is horrible and is usually preventable. It is up to you.

FIRST AID FOR GOATS

Goats almost always injure themselves or first show signs of illness on Sundays or in the middle of the night, when its almost impossible to get a veterinarian. Hopefully, some of the following suggestions will eliminate the need to call the vet and others will at least postpone the immediate need.

One of the most common emergencies is the "dog attack." Actually, anything causing puncture wounds, small cuts, or large lacerations will fit into this category. Most puncture wounds should not be sutured, but should be cleaned and treated with local antibiotics (furacin or iodine). Make sure that the animals tetanus vaccinations are up-to-date since punctures are an ideal place for the tetanus organism to grow. Most punctures should not be allowed to close over immediately but should be flushed out for several days to eliminate any possibility of infection.

Small cuts are only of major concern if they are deep and cut into major blood vessels, tendons, ligaments, or joints. Large or deep lacerations will almost always need veterinary care. If a wound is very dirty, rinse it with lots of clean water but without much scrubbing or manipulation. If it is bleeding excessively, apply a clean pressure bandage. An old clean sheet torn into strips will work well for bandage material, and a sanitary napkin is ideal for absorbing blood and applying even pressure. Do not apply "blue lotion" or other antiseptics when you know the vet will be seeing the goat soon.

Intestinal "emergencies" are the next most common. If your animal is not eating but is still alert and able to walk around the pen, try my Milk of Magnesia and yogurt remedy first. If you don't see a response within 12-24 hours, or if the animal worsens, call the vet. If the animal is severely bloated and uncomfortable, one dose of Milk of Magnesia should relieve the problem within 1 hour - or call the vet. Do not use this treatment if the animal is choking (coughing, gagging, and drooling profusely - also bloated).

Milk of Magnesia remedy:
      Milk of Magnesia (any flavor)
          - 2 oz. every 4 hours
      Yogurt - 1/2 cup once daily

Diarrhea, unless it is bloody, can also be treated with the Milk of Magnesia remedy. But constipation is easily confused with straining to urinate and home treatments are not usually of help. Call the vet!

A First Aid Kit for Traveling

1. Thermometer - and remember that the goat's normal

temperature is 102.5 - 104°F

2. Iodine (2%) - good antiseptic for scrapes, hooves trimmed too short, etc.

3. Lactobacillus acidophilus capsules - for animals that are slightly off feed. It will help prevent major rumen upsets.

4. Milk of magnesia - for rumen upsets, diarrhea, or any intestinal upset in animals over 3 months of age. Give 1 oz. per 30 pounds.

5. Molasses - use at home prior to traveling in drinking water, and continue to use in water on the road. The animal is less likely to go "off water." Also use for early case of pregnancy toxemia.

6. Turkey baster (plastic!) - use to administer #3,4, and 5.

7. Antibiotic - give to animals which have a fever. Do NOT give to a normal goat with a clear nasal discharge. (Tylan® or Liquamycin-LA® do not have to be refrigerated; Penicillin compounds should be refrigerated)

8. Epinepherine - mandatory to have available if you are going to give any other drugs, especially by injection. Drug reactions do not wait while you search out a bottle of epinepherine.

9. Syringes - use to administer #7 and 8.

10. Antibiotic ointment (Furacin®, Nolvasan®) - for minor cuts or scrapes.

11. Antibiotic eye ointment (Neosporin® ophthalmic) - for minor eye irritations that don't respond to a clear water wash. Check with vet as soon as you get home if the irritation doesn't clear up immediately.

12. Aspirin tablets (or dipyrone injection) - use to combat high fever - may use in conjunction with antibiotic. Give aspirin: 5 tablets (325 mg each) per 50

pounds twice daily.

13. Bandage material (gauze and tape) - may use for leg injuries on goats - or various cuts on humans.

14. Louse powder - smart to take as a precaution - dust all animals before returning to pens at home.

15. Fly spray - nice addition during hot, summer trips.

SURGICAL
MANAGEMENT

ANESTHESIA

Pygmy goats do require a little special handling during anesthesia. They are ruminants and care must be taken with heavy tranquilization or anesthesia to avoid bloating and/or regurgitation. Proper preparation and management will minimize complications.

Pygmy goats are usually docile enough to allow many procedures to be performed with no chemical restraint. If they aren't docile enough, they are small enough to be physically restrained during non-painful procedures. A discussion of various methods of restraint begins on page 109.

When physical restraint fails, or causes too much mental trauma to the goat, some form of chemical restraint must be used. It is important to remember that drug dosages are usually reported in ranges. Since animals all respond differently to drugs, the dose that achieves a perfect level of sedation in one goat may be way too much or too little for the next goat. I recommend starting with the lowest recommended dose. You can give a second dose if needed, but it is much more inconvenient to deal with an animal that is down longer and deeper than you expected.

SEDATION/TRANQUILIZATION
Definition: The act or process of calming.

Mild sedation can be achieved with acepromazine at 0.05-0.1 mg/lb subcutaneously. The goal is to cause the goat to be less apprehensive but to remain standing.

205

Acepromazine will work most reliably when given to a calm goat. Give the drug before fighting with the goat and then leave it along for 15-20 minutes to allow the drug to take effect. If the goat is anxious and stressed when the drug is administered, it will not achieve the same degree of sedation.

A low dose of xylazine (Rompun®) can also be used for sedation. 0.025-0.05 mg/lb intramuscularly will usually achieve an adequate level of sedation in a goat that can still stand and walk.

ANESTHESIA
Definition: Loss of feeling or sensation, especially loss of tactile sensibility, though the term is used for loss of any of the other senses.
General anesthesia: the production of complete unconsciousness, muscular relaxation and absence of pain sensation.

Local
Many minor procedures can be performed on goats with local anesthesia. Sedation may be required in some animals along with the local. Teat and horn surgeries are especially amenable to local blocks. A ring of lidocaine around the base of the teat will allow surgery on any part of the teat.

Intramuscular
Xylazine can be given at 0.1 mg/lb intramuscularly. This dose will usually cause the goat to become recumbent. Xylazine has minimal analgesic effect; it does not stop pain. Procedures done with xylazine anesthesia may need additional local anesthesia or other

drugs for pain control. The effects of xylazine can be partially reversed by yohimbine (Yobine®) given intravenously at 0.05 mg/lb.

Ketamine/xylazine – a combination of ketamine and xylazine will give surgical anesthesia with analgesia. The intramuscular dose is 0.1 mg/lb of xylazine followed in 3-5 minutes with 3 mg/lb of ketamine. This dose should give 15-30 minutes of surgical anesthesia. Additional doses of ketamine can be given to prolong anesthesia.

Intravenous

Xylazine can be given intravenously. It will have a more profound effect but for a shorter period of time than the same dose given IM. 0.05mg/lb should be a good dose when given IV.

A combination of ketamine and diazepam (Valium®) will provide excellent anesthesia for shorter procedures when inhalant drugs are not available. It can also be used to allow intubation for maintenance on inhalant anesthesia. These two drugs can be mixed in the same syringe at a 1: 1 ratio. 1 cc/ 20 lb of the mixture is given IV.

Inhalant

Isoflurane is the ideal inhalant drug for use in goats. Halothane can also be used, but an occasional goat may develop liver damage even with one isolated procedure.

The goat can be masked down or given Ketamine / valium prior to intubation. Intubation is relatively easy when done by "feel." With the goat in left lateral recumbency, grasp the larynx in the left hand. Insert the endotracheal tube in the mouth and push it just past the larynx. Pull it out just far enough to open the epiglottis then rotate the tube 45 and reinsert it into the larynx. By stabilizing the larynx with the left hand, it helps manipulate the tube into the right channel. Be sure to inflate the cuff to avoid aspiration pneumonia. It is advisable to place a second tracheal tube in the esophagus and inflate the cuff. This will allow the goat to expel rumen gas but helps protect against accidental aspiration.

With all methods of anesthesia, handling during recovery is very important. Goats must be maintained in sternal recumbency as soon as surgery is over and any tubes have been pulled. This position will allow normal eructation and prevent problems with bloat. Leaving a goat in lateral recumbency may cause fatal bloat. Prop the goat up against a fence or between two bales of hay.

Castration

The law of averages says that approximately 50% of your kid crop will be bucks. Since we're all breeding and looking for the "perfect" buck, it is nice to be able to choose from several. But since we're also being very careful not to keep poor quality bucks, we're all left with a lot of bucks to castrate. What is the best way to accomplish this task?

Castrating, or wethering, a buck kid can be done by at least 3 different methods. Each method has it's own pros and cons and you must use the method that best suits your management practices.

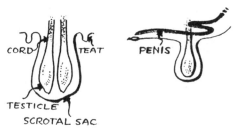

Figure 72. Scrotal anatomy - front and side view

1. Elastration, or banding, is the safest and easiest castration method for use by the average breeder. An elastrator band looks like a small cheerio-sized green or white rubber band. It is stretched by an elastrator instrument and applied around the base of the scrotum. Once released from the stretcher, it constricts the blood vessels leading to the scrotum and it's contents which will wither up and fall off within 3 to 6 weeks. Several precautions must be taken for safe use of bands. The bands should be carefully disinfected with alcohol prior

to use.   Care should be taken to see that BOTH testicles remain in the scrotal sac when the band is released (see **Figure 73**a). If one testicle slips up above the band, the animal will continue to smell and act like a buck, even though he will probably be sterile. Care should also be taken to keep the sac loose when applying the band. If the scrotum is pulled away from the body and the skin is held taut, the band will cut through the skin too quickly and leave a defect in the skin which will be slow to heal. If the scrotum is pulled tight and the band is applied very close to the body, the penis may be included in the band (see **Figure 73**b), causing painful rupture of the bladder and death of the kid in just a few days. And since the band takes several weeks to slowly cut through the base of the scrotum, it is extremely important to give tetanus antitoxin when the band is applied and to follow up in two weeks with tetanus toxoid.   Bands may cause severe discomfort to some kids for up to 4-6 hours, while other kids show no ill effects at all.   The end result is a kid with no scrotal sac at all (see **Figure 73**c). While this method of castration sounds scary, it is quick and relatively easy to do once you learn to do it properly.

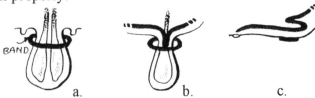

a.   b.   c.

Figure 73. a. proper application. b. improper application. c. final appearance.

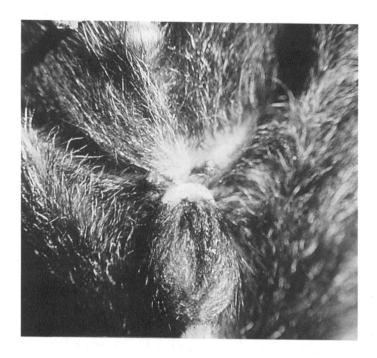

Figure 74. Elastrator band in place

2. The burdizzo is used by many veterinarians to crush the cords leading to the testicles. This kills the testicular tissue with no external wounds on the goat and little or no chance of tetanus or fly-strike. This procedure is usually done on kids six weeks of age or older since the burdizzo instrument is not small. A local anesthetic is usually injected in the cord to limit the pain of the procedure. Each cord is held to the side of the scrotum and crushed separately (see **Figure 75**a). Care is taken not to crush the center of the scrotum. A proper end result will have a shrunken scrotum with two small, firm, non-functional testicles inside (see **Figure 75**b). This method of castration is less than perfect and sometimes one cord will not be properly

211

crushed and will require a second treatment.

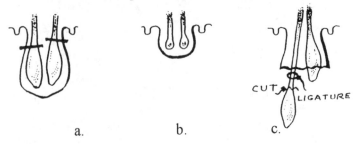

a.   b.   c.

Figure 75. a. proper burdizzo crimping b. final
result from burdizzo. c. surgical ligation.

3. Surgical castration is the "guaranteed" method. The
bottom one-third of the scrotum is incised or removed.
Each testicle is exposed separately and the cord either
ligated and cut (see **Figure 75c**), crushed and cut with
an emasculator, or scraped and cut with a knife. The
scrotum is left open for drainage and it will shrink and
heal quickly if all goes well. This method is most
humanely performed with anesthesia, but a skilled
operator can castrate an awake buckling with minimal
after effects of the stress. Fly control measures should
be taken until the scrotal wound has healed.

Pick the method of castration that works best for you.
Or take your excess bucklings to your veterinarian for
castration. Don't keep bucks because of a lack of
bravery to perform castration.

CRYPTORCHIDISM

Cryptorchidism does occur in the pygmy goat and
necessitates a veterinarian's involvement in castration.
With monorchidism, breeders should NOT remove the
descended testicle, so the veterinarian who must search

for the undescended gonad will know which side to search. Most often, the testicle will be in the inguinal region; it will have descended through the inguinal canal but not fully into the scrotum. Removal is straight forward with incision over the testicle, ligation and closure of the subcutis and skin. If the testicle is retained in the abdomen, a paramedian incision is made beside the sheath and entry is made into the abdominal cavity either through the linea alba or by muscle splitting. A "routine" search can be made for the retained testicle.

Ovariectomy

Goats can be quite obnoxious and noisy when in season, but this behavior has not created a trend toward spaying females. Cystic ovaries, on the other hand, cause health problems as well as continual obnoxious behavior. If medical treatment doesn't cure the problem then surgery is indicated. A ventral midline incision is made just cranial to the udder. The ovaries lie just cranial to the pelvis and dorsal to the urinary bladder. Ovaries with large cysts can be easily found by palpation. Exteriorize the ovary and ligate with the appropriate size absorbable suture material. Close the abdominal wall with #1 or #2 suture and close the skin as usual.

Wattle Removal

Figure 76. Wattles

Wattles are the remnants of gill slits from embryonic development. They appear as dangling pieces of skin on the neck (see **Figure 76**), or occasionally up under the ears. They have no effect on the health of the goat and no effect on the showing conformation of the animal. Sometimes they are not symmetrically placed on the goat and may cause an unbalanced look. The herdsman may want to remove those wattles.

The procedure is best done when the kid is just a few weeks old. Most wattles have a narrow "neck" near the junction with the body. Cutting right at that neck with a sharp pair of scissors will usually remove the wattle with little or no hemorrhage. A styptic pencil will stop

215

any minimal bleeding. If the wattle has a larger base or is being removed from an older animal, a local anesthetic can be used and a stitch will probably be required to close the wound.

Wattle cysts (also called branchial cleft cysts) are occasionally associated with wattles. This cyst appears as a small pea-sized to grape-sized mass at the base of the wattle. These cysts rarely cause any trouble and are best left alone. Do not tap the fluid out; you risk introducing infection and the cyst will merely refill. If a cyst does become infected, it can be surgically removed.

Mastectomy

Goats may develop tumors, severe mastitis, or abscesses in the udder that make a mastectomy the treatment of choice. An occasional goat will continue to fill her udder even after surgical removal of cystic ovaries. If the owner is not willing to milk her regularly, removal of the udder is indicated.

The goat is positioned in dorsal recumbency. An incision is made around the base of the udder leaving enough skin to close the wound. There are 3 large vessels to be found and ligated on each side. The mammary vein at the cranial border, the external pudic artery and vein near the inguinal ring, and the perineal artery and vein at the caudal border. Drains are usually required to assure good drainage in the early healing phase.

A mastectomy can be unilateral or bilateral. The halves of the udder can be separated on the midline and the good half left intact on the goat. Careful planning of skin flaps will allow a cosmetically pleasing closure of the wound.

Remember that the mammary gland is not necessary for reproduction. A goat can be bred and have kids following a complete mastectomy. She can NOT feed those kids so the herdsman must bottle feed. She can care for her kids otherwise. She can watch over them, help keep them warm, and teach them about being goats. Nursing is only one small part of "mothering".

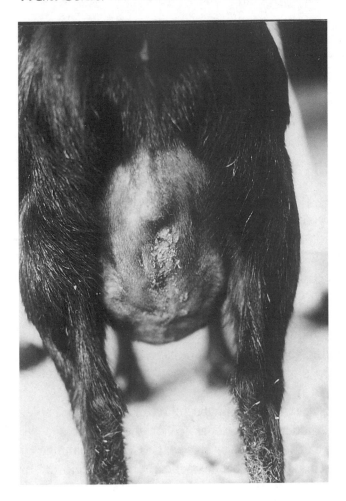

Figure 77. Post-mastectomy

Caesarian Section

A Caesarian section can be performed with a ventral midline or flank incision depending on the preference of the surgeon. The ventral midline approach is more likely to herniate or dehisce due to the weight of abdominal viscera on the incision. Flank incisions will leave visible scars that may affect the goats show career. The horns of the uterus can be exteriorized through either incision. The uterine incision should try to avoid the cotyledons. An inverting Lembert suture pattern should be used to close the uterus. Abdominal closure is routine. Keep the doe confined for 10-14 days after surgery.

Ideal anesthesia for a Caesarian section is isoflurane or halothane by mask followed by intubation. This will cause minimal depression of the kids and allow quick recovery of the doe.

If the surgery must be performed in the field, ketamine and xylazine can be used to achieve recumbency and a local line block will also be required. Once the kids are delivered, more ketamine can be given to deepen the level anesthesia for uterine and abdominal closure. The kids will be slower to recover from the effects of the anesthesia. Keep stimulating them until they are alert.

Debleating

Noisy pet goats can cause problems with human neighbors. Goats are smart enough to learn that screaming early in the morning gets them fed. Bad habits can be hard to break. If training methods fail, the goat can be debleated. It is preferable to postpone this surgery until the goat is close to a year old so that most of it's growing is done. Ketamine/xylazine anesthesia is used. A ventral midline incision is mode over the larynx. Incise through the larynx beginning at the cricoid cartilage and extend to the first tracheal ring. The vocal cords are small flaps of tissue just caudal to the arytenoid cartilages. Do not remove the arytenoid cartilages. Grasp the cord with a forceps and cut the entire cord out. Repeat this procedure with the other cord. There will be some hemorrhage; keep clots out of the trachea. Close the larynx with catgut. Observe for an airtight closure. Close the skin routinely. Observe carefully for 24 hours postsurgically for any subcutaneous emphysema. Resuturing may be necessary if any emphysema is noted.

Dehorning

Horns or no horns - the debate will go on as long as goats grow horns. We will avoid the debate. If you wish to have animals without horns, you disbud your baby animals. Unfortunately, sometimes the disbudding job is less than ideal and an ugly scur appears as the horns attempts to grow anyway. And, occasionally you will wish to remove the horns from an adult animal that you weren't able to disbud as a baby.

I dehorn mature animals during our non-fly season - usually from November through March. I refuse to do any animals during the fly season because even the most careful management has left me cleaning out maggots, etc. from the horn holes. Mature bucks will be left with a hole 1-2 inches in diameter into the sinuses which will take 2-4 months to totally heal. They may look and smell bad during that time, but most of the animals seem to take it right in stride and rarely miss more than one meal.

I tranquilize the animal with xylazine (dose = .05 mg/lb given IM) and then put in a local horn block. I use .5 cc of 2% xylocaine at each of four sites for a total of 2 cc of xylocaine per goat. One injection is made half way between the rear edge of the horn base and the lateral canthus of the eye - this injection is placed 1/8 to 1/4 inch deep. The other injection is made half way between the front edge of the horn base and the medial canthus of the eye - this injection is made quite superficial. Similar sites are used on the other horn.

221

The bases of the horns are shaved prior to the local anesthetic injections. The areas are scrubbed again while the local has a chance to work. Then I cut through the skin behind the horn with a scalpel blade to mark the area to "set" the gigli wire. The horn is removed with the gigli wire trying to include 1/4-1/2 inch of skin around all edges. An assistant helps hold the horn loosely in place to limit the hemorrhage. I then cauterize the bleeders with a soldering iron. I also place a cotton ball loosely in the top of the opening that is left into the sinus to help limit the air circulation through that sinus. As long as the cotton is not packed in tightly, it does not require removal at a later date, but it will slough off naturally with the scabs, discharge and serous debris that occurs. I do not like to bandage the head following dehorning but would rather allow good air circulation to the area.

Aftercare is minimal. Fly control is very important. The only times I have really had any problems following dehorning have been in very rainy weather if the animal did not get into shelter and got rain into his horn holes. Then I see a high fever and have to treat the animal with antibiotics. If the cotton ball packing comes out in the first few days, I recommend gently replacing it with a new one.

This same procedure works for all horns that are too big to be properly disbudded and for scurs. However, since scurs are deformed horns, it is often difficult to find the proper landmarks for a good local anesthesia. Very small horns may have no hole into the sinus and will not require any cotton packing.

APPENDIX

DRUG DOSES

DRUG	Route	Dosage
Acepromazine injection	sq	0.05-0.1 mg/lb
Ampicillin injection	im	5 mg/lb sid
Amprolium	po	mix 3 oz/pint of water/give 1 oz/100 lb sid x 5 days/then give 0.5 oz/100 lb sid x 21 days
Aspirin	po	0.5 gr/lb bid
Atropine injection	im/sq/iv	1 cc/25 lb (1/120gr/cc)
Azium tablets	po	5-20 mg sid
Baking soda	po	1 tsp/20 lb
Bo-Se injection	im/sq	1 cc/40 lb twice yearly
Combiotic®	sq	1 cc/20 lb bid
Copper Sulfate	foot dip	
Cystorelin®	im/iv	0.25-0.5 cc (15-25 mcg)
Decoquinate (Deccox®)	po	0.25 mg/lb (0.5mg/kg) x 28 days
Diazepam (Valium®)	im/sq/iv	2-5 mg/10 lb
Doxapram (Dopram®)	im/iv	0.2 mg/lb 2-10 drops under tongue for newborn

Drug	Route	Dose
Epinepherine (Adrenalin®)	im/sq/iv	0.1-2 cc of 1:1000 solution kids - 0.1-0.25cc IM adults - 1-2cc IM
Erythromycin (Erythro 200®)	im	0.5 cc/100 lb sid
Fenbendazole (Panacur®)	po	2.5 mg/lb(nematodes), 7 mg/lb (cestodes)
Flocillin®	sq	2cc/150 lb
Flunixin (Banamine®)	im/sq/po	0.5 mg/lb (0.1cc/10#)
FSH-P®	im/sq/iv	5-25 mg
Formaldehyde	foot dip	
Furosemide (Lasix®)	po	1-2 mg/lb sid
Furosemide injection	im/sq	2 mg/lb
Gentamycin injection	sq	2 mg/lb
Injacom	im	2.5-3.5 cc
Iodine	po	5 drop Lugol/1 oz water sid x 5 days
Ivermectin (Ivomec®) inj	sq/po	0.1 mg/lb
Ketamine	im	5 mg/lb
Levamisole (Tramisol®)	po	3.5 mg/lb (1 bolus/50 lb)
Lidocaine injection		2-3 cc/10 lbs is toxic
Liquamycin LA-200®	im	1cc/22 lb eod (or 1cc/40 lb sid)

Drug	Route	Dosage
Lixotinic®	po	0.5-1 oz
Mebendazole (Telmin®)	po	4 mg/lb
Milk of magnesia	po	1 oz/30 lb
Oxfendazole (Benzelmin®)	po	2.5 mg/lb (1/10 packet pellets/100 lbs)
Oxytetracycline (Liquamycin®)	iv	1-2 cc 50 lb (3-5 mg/lb bid)
Oxytocin	im/sq	30-50 units
Penicillin G, Procaine	im/sq	3-10,000 units/lb sid
Penicillin G, Sodium	iv	5-10,000 units/lb qid
PLH®	iv/sq	2.5 mg (0.5 cc)
Praziquantel (Droncit®)	po	4.5 mg/lb
Progesterone acetate	im	0.5 mg/lb
Propylene glycol	po	2 oz bid
Pyrantel (Nemex®)	po	10 mg/lb
Sulfadimethoxine (Albon®)		
injection	sq	3 cc/50 lb then 1.5 cc/50 lb for 5 days
oral suspension	po	1 tsp/10 lb then 0.5 tsp/10 lb for 5 days
tablets	po	250 mg/10 lb then 125 mg/10 lb for 5 days
Sulfamethazine	iv/po	100 mg/lb
Thiabendazole	po	20-30 mg/lb (1 bolus/100 lb)

Drug	Route	Dosage
Thiamine injection	im/iv	500 mg (1/2 im and 1/2 iv)(10mg/kg qid x 24h)
Tribressin® injection	iv/sq	2 cc/100 lb/day (48%)
Tribressin tab (480 mg)	po	1 tab/40 lb sid
Triple Sulfa	iv/po	100 mg/lb sid
Tylocine injection	im	1-2 mg/lb sid or bid
Vetalog injection	im/sq	1-2 cc
Vitamin B$_{12}$ (Cyanocobolamin)	im	100 mcg +
Vitamin E (alpha tocopherol)	im/po	300 iu/kid im-700 iu/doe im
		50 mg/kid/day orally
Xylazine (Rompun®)	im/iv	0.05-0.1 mg/lb
Yogurt	po	1 cup / day / 100-150 pounds
Yohimbine (Yobine®)	iv	0.05 mg/lb
Zinc sulfate	po	250-500 mg/day X 4 wk

routes:
    iv - intravenous
    im - intramuscular
    sq - subcutaneous
    po - orally (per os)

dosage:
    sid - once daily
    bid - twice daily
    tid - three times daily
    qid - four times daily
    eod - every other day

measurements   mg - milligrams
    iu - international units
    cc - cubic centimeters (milliliters)

GLOSSARY

acute - having a short and relatively severe course

anaerobic - growing only in the absence of molecular oxygen

anaphylaxis - an exaggerated reaction to a foreign substance, an antigen-antibody reaction

anorexia - lack or loss of the appetite for food

anthelmintic - a remedy for worms

auscultation - the act of listening for sounds within the body

brachydont teeth - low crowned or simple teeth consisting of a crown, neck and root

buck - male goat

caruncles - small fleshy eminence of the uterine wall

cestode - any tapeworm or platyhelminth (flatworm)

chronic - long continued; not acute

cotyledons - any one of the subdivisions of the uterine surface of a discoidal placenta

cryptorchidism - failure of the testes to descend into the scrotum

deciduous - not permanent, but cast off at maturity

debride - to remove foreign matter and devitalized tissue

dehiscence - the process of separation of all the layers of an incision

doe - female goat

exogenous - developed or originating outside the organism.

freshening - parturition in the goat

hypsodont teeth - high crowned or ever-growing teeth having no distinct neck

IU - International Units

leucocytosis - an increased number of white blood cells

monorchidism - condition of only one testes being descended into the scrotum

necrotic - pertaining to the death of a cell or group of cells which is in contact with living tissue

Nematoda - a class of helminths, roundworms or threadworms

nematode - endoparasite or species belonging to the Nematoda.

parturition - the act or process of giving birth

pathogenic - giving origin to disease

phagocytic cells - cells that ingest microorganisms or other cells and foreign particles

Protozoa - the lowest division of the animal kingdom, including unicellular organisms.

protozoan - an organism belonging to the Protozoa

reduce (a prolapse) - to restore to the normal place or relation of parts.

repel - to push back a fetus until the head and limbs can be properly placed for normal delivery.

visceral - pertaining to any large interior organ in any one of the three great cavities of the body, especially the abdomen

wether - castrated male goat

REFERENCES

1. Squires, Dr. Janine: Parasitology of Small Ruminants. Dairy Goat Journal, Oct. 86.
2. Blood, Radostits, and Henderson: Veterinary Medicine. Bailliere Tindall. Great Britain. 6th Edition.
3. Howard: Current Veterinary Therapy for Food Animal Practice. W.B. Saunders Company. 1981.
4. Oehme, Prier: Textbook of Large Animal Surgery. Williams and Wilkins. Baltimore, MD. 1976.
5. Smith, Mary C. and Sherman, David M.: Goat Medicine. Lea & Febiger. 1994.
6. Pygmy Goat Memo, Vol.1, No.2, Summer 1976, NPGA, Menden, MA 01756. With permission.
7. Pygmy Goat Memo, Vol.X, No.3, Fall 1985, NPGA, Menden, MA 01756. With permission.

LIST OF FIGURES

INDEX

All Publishing Company
10951 Meads Avenue
Orange, CA 92869

ORDER FORM

Name:_____

Address:_____

City:_____State:____Zip:_____

Please send:

_____ copies of **Pygmy Goats, Management and Veterinary Care** @ $19.95

_____ copies of The Illustrated Standard of the Pygmy Goat @ $9.95

_____ copies of Pot-Bellied Pet Pigs, Mini-Pig Care and Trainning, @ $9.95

_____ copies of Veterinary Care of Pot-Bellied Pet Pigs @ $20.00

Total for book(s)	
Sales tax @7.75% (Calif. only)	
Shipping and handling	
Total amount enclosed	

Shipping is $5.00 for the first book and $2.00 for each additional copy. Visa and MasterCard accepted. Prices are subject to change without notice. Please write for quantity discount rates.